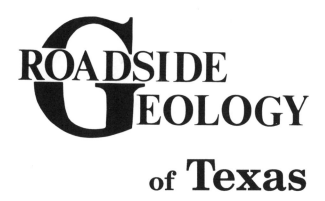

ROADSIDE GEOLOGY

of Texas

Darwin Spearing

MOUNTAIN PRESS PUBLISHING COMPANY
MISSOULA, MONTANA

Roadside Geology is a registered trademark
of Mountain Press Publishing Company.

Copyright © 1991
Darwin R. Spearing

Thirteenth Printing, January 2013

Library of Congress Cataloging-in-Publication Data

Spearing, Darwin.
  Roadside geology of Texas / Darwin Spearing.
    p. cm. — (Roadside geology series.)
  Rev. ed. of: Roadside geology of Texas / Robert A. Sheldon. 1979
  Includes bibliographical references and index.
  ISBN 978-0-87842-265-4
  1. Geology—Texas—Guide-books.   2. Texas—Description and
travel—1981—Guide-books.   I. Sheldon, Robert A.   Roadside
geology of Texas.   II. Title.   III. Series.
QE167.S53   1991   90-45051
557.64—dc20   CIP

PRINTED IN THE U.S.A.

MP Mountain Press
PUBLISHING COMPANY
P.O. Box 2399 · Missoula, MT 59806 · 406-728-1900
800-234-5308 · info@mtnpress.com
www.mountain-press.com

FOR SUE
Who shared a sparkling morning atop Enchanted Rock and
Thanksgiving dinner at an east Texas truck stop.

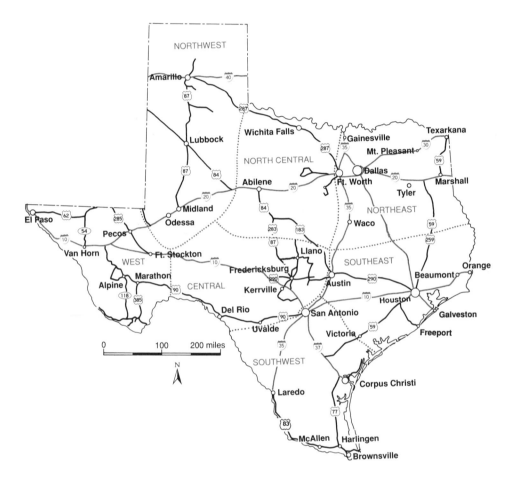

Roads and sections of the Roadside Geology of Texas.

Table of Contents

Publisher's Note

Back in 1979, we at Mountain Press published the first version of the *Roadside Geology of Texas*, written by the late Robert Sheldon. Most people think rocks are more or less permanent, that they last practically forever. But geologists know that they seem to change from one year to the next as new ideas emerge. Books about geology need to change, too, to keep up with the rocks.

We let the original *Roadside Geology of Texas* go out of print some years ago because it was too far behind the rocks. We then began looking for a new author to write a new book under the old title. We finally found Darwin Spearing, a distinguished geologist who lives and works in Texas.

So here it is, a completely new book by a new author published under the old title. It provides a completely new look at the old rocks of Texas, which don't look at all the way they did back in 1979. We are proud of this new book, and think you will enjoy it, too. Join Darwin Spearing for a tour of the fascinating rocks of Texas. Let him tell you what they are, how they got there, and what they mean. He tells their story with occasional wit, unflagging enthusiasm, and obvious authority.

Granite headstone—pure Texas.

Preface

HOW TO USE THIS BOOK

Texas is divided in this book into seven sections which represent relatively cohesive units of similar geology and, of course, geography. Each section begins with a map that shows the roads and geology, followed by a general written introduction to what you might expect to see as you drive the roads in the section. An attempt is made in each introductory review to set the geologic stage by describing how local rocks fit into the overall geological development through time. Before setting out on a trip across the state, determine which roads you will drive, find them in the appropriate section or sections, and study the map and read the introduction. The individual road descriptions and guides will then make a lot more sense.

About ten percent of the paved roads in Texas are described in this book, or about 7,500 miles of over 74,000 total miles of paved roads in the state. Descriptions are included for all the Interstate roads crossing Texas, as well as selected main highways that connect major towns or points of interest, such as National Parks, State Parks, and significant geological or popular recreation areas.

The purpose is to describe and explain the geology of Texas as seen from your car window while traveling to and fro across the state. This does not mean the best sites for geologic investigation are necessarily covered; in fact in some areas quite the contrary is true. In east Texas, for example, the best geologic sections are commonly seen in river banks and railroad cuts, and geologists visit these sites regularly. However, the intent of this book is to tell you what you're seeing as you drive common roads, and it would be inappropriate to direct travelers miles out of their way, down gravel roads and up creek beds to see a special geologic section. Guidebooks are available to these special sites, and sources for such guides are listed in the back of the book for readers interested in pursuing these ventures. Remember also that the best geology may not be seen along major roads and freeways for several other reasons. First, highway engineers tend to avoid stream crossings and rough terrain when they plan their roads, exactly where the best geology is to be viewed. Secondly, Interstate and major highway roadcuts are commonly graded and grassed, especially in the eastern half of the state, so once-marvelous cuts, originally exposed when the roadway was built, are now mostly covered by grass and wildflowers.

If you travel on roads not specifically covered by this guide, refer to a state highway roadmap and compare it to the geologic map at the head of the section in which you're driving. You will then be able to interpret, on your own, much of the roadside geology on the unguided road. You may also find it helpful to order the "Geological Highway Map of Texas", from the American Association of Petroleum Geologists, Post Office Box 979, Tulsa, Oklahoma 74101.

Special sections are included for the National Parks and many of the State Parks and recreation areas. Readers who may seek additional information on the parks will find reference to more specific texts, guides and pamphlets at the end of these sections.

Geologic word use and professional jargon have been minimized in this book, but it is impossible to present a scientific subject such as geology without using at least some geologic words. If a term is unfamiliar, please look it up in the glossary provided at the end of the book.

This book is not written for geologists. It is intended for intelligent, non-geologists who want to know something about the geology of Texas as seen from their car. To cover the entire state in a book of this size means that many of the details geologists normally look at when they study a section of road have been purposefully omitted for the sake of space. Therefore, this is not a guide to every mile of geology to be seen between points A and B. Only the main features and highlights, which hopefully illustrate the principal geologic ideas which are exemplified along a particular road, are included. Detailed roadguides are available for many roads in the state; if you have a particular interest in an area and want to know more, please refer to the section on "Where to Get Maps and Information" at the back of the book.

One of the rather consistent themes throughout the text is geologic time. Geologic maps show areas where "Cretaceous" rocks, for example, are dominant in a certain area. Readers might ask, why the attention to age of rocks? The geologic history of the earth, like a giant jigsaw puzzle, can only be understood by assembling widely separated pieces. If you know what happened in Eocene time in East Texas, (sandstone, shale, deltas, rivers, coals, swamps) then travel to west Texas and see Eocene rocks again (lava flows, ash falls, mountain building) a whole scenic panorama for that period of time begins to develop in your mind for the entire state of Texas. By knowing the age, comparisons can be made from place to place to help create a complete environmental picture. The unfolding of ancient environments is one of the most exciting contributions geology has to offer to the human mind. It is the only true time-travel experience we will ever have, science-fiction notwithstanding!

The other major theme is how the earth works, how the common processes going on around us every day help to shape and mold the surface of the earth. The earth's surface is dynamic and ever- changing, and mountains, for example, are not permanent monoliths on the earth surface. Of course, they seem to be when compared to spans of human life, but not in terms of geologic time measured in millions, even billions of years. Being able to look at mountains in West Texas or caverns in Central Texas, or canyons in the Panhandle and envision the forces and changes - the *dynamics* - of the landscape you are seeing is a real achievement of human understanding, and one hopefully you will learn, or further appreciate, as you study the *Roadside Geology of Texas*.

Acknowledgements

Sincere thanks are extended to a number a people who directly contributed to this book. Billy P. Baker reviewed and identified photographs of dinosaur tracks. Virgil E. Barnes, Patricia Wood Dickerson, Samuel P. Ellison, Jr., Thomas E. Ewing, Earle F. McBride, Robert A. Morton, Austin A. Sartin, Gerald E. Schultz, and E. G. Wermund ably reviewed and constructively criticized text and illustration drafts, measureably improving both. The time and effort contributed by these busy geologists are greatly appreciated and sincerely acknowledged. Acknowledgement of these reviewers does not constitute an endorsement of this book by them, however; the author is solely responsible for the content.

Thanks are also due to the people at Mountain Press, especially Dave Flaccus, for their able assistance. David Alt and Donald Hyndman edited the entire manuscript and made many helpful suggestions.

Darwin R. Spearing

GEOLOGIC TIME SCALE

	O TODAY
CENOZOIC	66 MILLIONS OF
MESOZOIC	YEARS AGO
	245
PALEOZOIC	
	570
	1000

If the geologic eras are scaled in actual years, this is how the geologic time scale appears.

P R E C A M B R I A N

2000

3000

Most of Earth's history is Precambrian

4000

4500

ERA	PERIOD	EPOCH	M.
			C
C E N O Z O I C	Quaternary	Recent	C
		Pleistocene	
			2
	Tertiary	Pliocene	
			5
		Miocene	
			2
		Oligocene	
			3
		Eocene	
			5
		Paleocene	
			6
M E S O Z O I C	Cretaceous		
			1
	Jurassic		
			2
	Triassic		
			2
P A L E O Z O I C	Permian		2
	Pennsylvanian		
			3
	Mississippian		
			3
	Devonian		
			4
	Silurian		
			4
	Ordovician		
			5
	Cambrian		
			5
PRE-CAMBRIAN	Proterozoic		
			2
	Archean		
			4

I
The Big Picture

GEOLOGIC TIME

There is nothing magic or overly complicated about geologic time. Geologists have simply divided up the geologic history of the earth into convenient packages, named the packages for ease of communication and called it the geologic time scale. The more-or-less official geologic time scale, used throughout the world, has been around for over 100 years now, and was developed quite some time before anybody had any idea of how long ago, in years, the Permian period, for example, actually occurred.

Thus, the geologic time scale you see on the adjacent page is based on relative time. For example, geologists originally recognized a group of rocks with a distinctive set of fossils that occurred in England, and then they saw similar rocks with similar fossils in Russia, and named the collective assemblage the "Permian" period. In other words, they interpreted that the two sets of similar rocks, though quite far apart, were deposited at the same time in the geologic past. But then these geologists or maybe other geologists, recognized this "Permian" package of rocks always lay under another set of rocks, distinctive in themselves, that had a different fossil assemblage. They named this second group of rocks "Triassic", and so the time scale developed with continued observation of rocks around the world. A stack of rocks representing the very oldest to the very youngest was recognized, named and ordered and by about 1900 geologists everywhere pretty well agreed on this relative age scale. Anyone with some basic knowledge could go out anywhere in the world, look at a set of rocks, study the fossil assemblage, and determine that these rocks were deposited in

"Permian" time, thereby fitting these new examples into the grand scheme of things, geologically speaking.

But nobody knew how old any of these rocks were! Geologists blithely talked about "Permian" rocks and could say they were older than the Triassic ones, and younger than the Devonian ones, but whether these Permian rocks were formulated 50 years ago or 50 thousand years ago or 50 million years ago or 500 million years ago, no one had any idea nor really any way of measuring the actual ages until the constancy of radioactive decay of certain elements, such as Uranium, gave the world atomic clocks early in the 20th Century. To say no geologist prior to 1900 had any idea about the age of the earth is somewhat a misstatement, because plenty of ideas floated around. Some people calculated how salty the ocean was and how long it would take rivers to pump enough salt into the sea to get the salt percent to where it is today, thereby estimating the age of the oceans. Others guessed about erosion rates and how long it would take to flatten a mountain range over time, and so on. And, for the most part, the 19th Century geologic fraternity believed, through these rather inexact measuring techniques, that the age of the earth was probably a few million years. It is difficult for us today to understand the immense psychic impact on the human mind early in the 20th Century when the constancy of radioactive decay was discovered, and the first age dates from the ensuing Uranium clocks turned up several hundred million years for Paleozoic rocks, and a few billion year dates for the really old Precambrian rocks! People were shocked, and the ensuing expansion of the human race's concept of the immensity of geologic time is a major scientific contribution of this Century.

Since then, radioactive age dates have been combined with the relative geologic time scale, so now geologists can recognize "Permian" rocks in the field, but also know these rocks were deposited between 245 and 286 million years ago. And, it probably is not surprising to know that a major effort has continued over the last 75 years to refine and upgrade the radioactive age dates, by measuring more and more samples.

Refer back to the geologic time scale as often as you feel the need—after awhile you'll probably have a lot of it memorized through use.

A MARCH THROUGH GEOLOGIC TIME

The accompanying maps and descriptions outline major events in Texas for each geologic period or epoch.

Rock symbols on the maps of the geologic epochs.

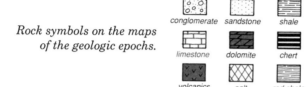

This section is designed to serve as a handy reference and you will probably thumb back to it frequently as you follow the road guides and find yourself trying to recall, for example, just what *did* happen in Jurassic time?

PRECAMBRIAN:
570 - 4,500 Million Years Ago

The geologic record begins in Texas a little over a billion years ago, when thick sequences of coarse, then fine sediment were dumped into an ancient sea bordering a continent. Eventually the continent collided with either another continent or an ocean margin in a plate tectonic event that buried, squeezed and heated the borderlands, including the sediment piles. The collision built mountains and created metamorphic schist and gneiss out of the deeply buried sediments and generated molten magmas which cooled to form granite bodies. Erosion then flattened this range to a table-top by Cambrian time. All these rocks are displayed in the Llano country of central Texas. Precambrian rocks are also seen in the Franklin Mountains near El Paso, and in west Texas ranges near Van Horn. Precambrian rocks have been reached by oil drilling over much of central and west Texas, though little is known about Precambrian rocks beneath the Gulf Coastal Plain because they are buried so deeply there.

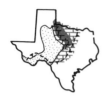

CAMBRIAN:
505 - 570 Million Years Ago

Shallow marine seas that transgressed across Texas in Cambrian time bordered the low, erosion-worn central core of the North American continent.

Sandy sediments were deposited at the margin of these seas from streams carrying their loads eroded from the low continental terrain to the north and west. Cambrian sandstones around the Llano uplift in central Texas and in the Marathon region of West Texas are examples of this sedimentation.

Farther from shore in clear marine water, dolomite and limestone accumulated from the shells of Cambrian organisms. Outcrops of Cambrian limestones are also found in the Llano uplift area of central Texas. Cambrian time is noteworthy because it represents the appearance, rather suddenly in the geologic record, of abundant fossils. Trilobites, brachiopods, sponges, snails, clams, and bryozoans were all present by the Cambrian time, whereas late Precambrian rocks display only rare fossils of algae and a few soft-bodied marine animals.

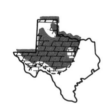

ORDOVICIAN:
438 - 505 Million Years Ago

Extensive dolomite and limestone deposits, with less extensive chert deposits, were laid down in shallow seas that covered Texas in Ordovician time. Remnant Ordovician outcrops are best seen in the Llano uplift northwest of Austin, and in the Marathon uplift and Franklin Mountains of West Texas. The Ellenburger and Maravillas formations appear along roadsides. The Ellenburger is especially noted for its gas production from great depths in basins of West Texas. Other Ordovician limestones yield brachiopods and corals on the top of the El Paso scenic drive, north of El Paso in the Franklin Mountains, while cephalopods and gastropods are found in Ellenburger outcrops north of Cherokee in the Llano uplift region. Maravillas limestone and chert beds are seen in the Marathon uplift around Marathon in West Texas.

SILURIAN:
408 - 438 Million Years Ago

Limestone, dolomite and chert rocks were laid down in shallow marine waters in West Texas during the Silurian period. Silurian limestones in the Franklin Mountains north of El Paso bear rare brachiopods and corals. The first primitive land plants appear in the Silurian, and the corals, first seen in late Ordovician time, exploded in numbers and types during the Silurian.

DEVONIAN:
360 - 408 Million Years Ago

Shale, sandstone, limestone, and chert of shallow marine origin characterize the rocks of Devonian age in Texas. Devonian limestone, sandstone and shale deposits are well known in the subsurface, and appear in outcrops in the central Texas Llano uplift and the Marathon uplift of West Texas.

Amphibians first appear in the Devonian, and this was the golden age for the development of fishes. On land, ferns, seedferns, and huge trees related to present-day horsetail rushes, developed in the Devonian period, but exploded in numbers during the Mississippian.

MISSISSIPPIAN:
320 - 360 Million Years Ago

Shallow marine seas still covered Texas in the Mississippian period wherein marine shales and limestones were deposited. On land were fern-filled forests, while in the seas brachiopods, bryozoans, trilobites and corals were common.

Mississippian shales and limestones are found around the Llano uplift, and folded, upended Mississippian rocks occur in roadcuts east of Marathon in West Texas.

PENNSYLVANIAN:
286 - 320 Million Years Ago

About 300 million years ago the Ouachita Mountain range rose to form a distinctive feature across Texas, as the then-North American continent collided with another continent to become part of the Pangaea supercontinent. To the west of the uplifted Ouachita Mountains, the crust sagged in response and several basins formed. Seas and sediments found their way into these basins.

Thick sections of Pennsylvanian-age marine limestones, along with shales and sandstones, underlie most of West Texas. The Cisco, Canyon, Strawn, and Bend are Pennsylvanian groups of rocks, arranged in descending order of age.

Pennsylvanian rocks are found in roadcuts north and east of Marathon, West Texas. From the Llano uplift northward to Jacksboro and Bowie lies a wide, inclined belt of Pennsylvanian rocks arranged in bands, oldest to youngest, from east to west. Marine fossil snails, clams, trilobites, bryozoans and ammonites are found in Strawn group rocks east of Mineral Wells, north central Texas. Canyon group snails, clams and crinoids are found along U.S. 377 southwest of Brownwood, central Texas. Cisco group brachiopods and clams occur northeast of Cisco, in north central Texas. Fossils of fusulinids - one-celled organisms that look like grains of wheat - are very common and characterize rocks of Pennsylvanian and Permian age. Marble Falls limestone can be seen along the river at Marble Falls on the east side of the Llano uplift.

Extensive forests of conifers, ferns, seedferns and horsetail trees in the Pennsylvanian period gave rise to coal deposits (in Pennsylvania, the state), and reptiles first roamed these forests at the end of the period. The Mississippian and Pennsylvanian periods are together called "Carboniferous" over much of the world, because of the abundant coal deposits laid down during this time.

PERMIAN:
245 - 286 Million Years Ago

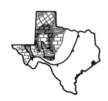

Permian rocks are a geologic delight in Texas! Colorful red beds at the edge of the high plains (Caprock Canyons State Park, Palo Duro Canyon) in the Panhandle are Permian, as are the magnificent reef limestones on El Capitan in Guadalupe National Park, as well as oil-rich limestones in the subsurface of West Texas. Great climatic changes occurred in the Permian as major groups of corals, bryozoans, brachiopods, fusulinids and crinoids became extinct, along with the entire trilobite tribe. But vertebrates such as the mammal-like reptiles (fin-backed *Dimetrodon* is an example) flourished. Land plants also changed, and the ferns, seedferns, and horsetail trees began to decline during the Permian. But, Plesiosaurs, lizards, thecodonts (precursors of dinosaurs and crocodiles) and therapsids (mammal-like reptiles) had their beginnings in the Permian period.

While marine reefs and banks prevailed in shallow marine waters of West Texas, near-shore evaporation flats in the Panhandle area produced deposits of bright red shales, and salt and gypsum deposits.

Permian rocks are at the surface in vast expanses of terrain in north Texas from the edge of the high plains eastward to Mineral Wells and Wichita Falls, southward past Abilene and to San Angelo. Tracts of Permian rocks extend northward from I-10 to Guadalupe National Park in West Texas and westward nearly to El Paso. There are dozens of formation names to describe various Permian-age rocks in Texas. Some worth mentioning here are the upper Permian Quartermaster formation (red sandstones, shales, evaporite minerals) in Palo Duro Canyon, and the Capitan, Goat Seep, Brushy Canyon and Bone Springs formations in Guadalupe National Park.

TRIASSIC:
208 - 245 Million Years Ago

By Triassic time, colorful shales and sandstones were still being deposited, although in more restricted areas of the Panhandle. The Dockum group is well exposed at Palo Duro Canyon and Caprock Canyons State parks and in the breaks along the Canadian River west of Amarillo, and a wide patch of Triassic rocks occurs east of Big Springs.

However, rumbles of change in the configuration of continents are seen in Texas as the supercontinent of Pangaea begins to split apart during the Triassic. The Gulf of Mexico begins to shudder open, and red shale, siltstone and sandstone are the first deposits to be shed into the down-warping southeast and east side of the Ouachita mountain line.

On land, ferns begin anew, cycads appear, while dinosaurs, crocodiles, ichthyosaurs and turtles arrive on the scene. In the sea, true corals begin, and there is a re-awakening of bryozoans.

JURASSIC:
144 - 208 Million Years Ago

During Jurassic time, the breakup of Pangaea began in earnest, the Rocky Mountains were rising and the Gulf of Mexico occupied the new gap between North and South America. At the beginning the Gulf was a shallow sea, not well connected to the other oceans, because it dried up often, creating vast salt pans. The Louann Salt, mother lode of salt domes in the Gulf Coast, was thus born. Limestones of the subsurface Smackover formation were deposited when deeper marine water conditions prevailed.

The dinosaurs were in full force by the Jurassic period, and the first kinds of flowering plants and early rodent-like mammals appeared.

Jurassic rocks are virtually absent at the surface in Texas, though Jurassic limestone, sandstone and shale beds can be seen along I-10 west of Sierra Blanca in westernmost Texas.

upper
Cretaceous

lower
Cretaceous

CRETACEOUS:
66 - 144 Million Years Ago

Texas Cretaceous rocks are fascinating! Lower Cretaceous rocks virtually blanket the center half of the state, and limestone cliffs, caverns, canyons, springs, abundant fossils, and dinosaur tracks are all part of the "Cretaceous scene." Upper Cretaceous rocks are found in a band from the Red River southward through Dallas/Fort Worth, to Austin, San Antonio and westward to Del Rio, and in Big Bend National Park. The interstate highway (I-10) from San Antonio to Fort Stockton runs entirely on Cretaceous rocks, and wonderful canyon outcrops and roadcuts are seen along the road.

The continents continued to pull apart in Cretaceous time. The Rocky Mountains underwent their major push (the Laramide orogeny) and shallow seas on the new continents' margins advanced and retreated repeatedly. Some sea advances filled the trough in front of the Rocky Mountains, creating a connecting seaway all the way from the Artic Ocean to the Gulf of Mexico. The shallow Cretaceous seas over Texas were filled with calcareous-shelled organisms, and thick deposits of limestone were laid down. On the sandy shorelines and mudflats of these seas dinosaurs roamed freely, leaving evidence of their passing in fantastic fossilized footprints and trackways all across Texas.

Many formation names are applied to describe the complex suite of Texas' Cretaceous rocks. The lower Cretaceous portion is divided into Washita, Fredricksburg, and Trinity groups from top to bottom, while upper Cretaceous rocks are assembled into Navarro, Taylor, Austin, Eagle Ford, and Woodbine groups, top to bottom.

Cretaceous rocks form the impressive cliffs at Santa Elena and Boquillas Canyons in Big Bend Park. The Hill Country around San Antonio and Kerrville is carved in Cretaceous rocks, as are the Colorado River Canyon north of Austin and the Devils River-Rio Grande Canyon west of Del Rio.

While marsupials and bats arose in the Cretaceous, and though dinosaurs dominated and flowering plants proliferated, the winds of major change blew for life on earth during the Cretaceous. At the close of Cretaceous time (66 million years ago), the dinosaurs and many of their relatives disappeared forever. But strangely, plant life marched across the Cretaceous boundary into the Tertiary period virtually unchanged, as did the birds. Though the picture of dinosaurs choking to death from the cataclysmic dust of a meteor impact is a commonly touted idea, it does not explain why flowering plants or birds survived the holocaust! More work needs to be done on the geologic causes of the so-called mass extinction at the end of the Cretaceous.

TERTIARY:
2 - 66 Million Years Ago
(Paleocene, Eocene, Oligocene, Miocene, Pliocene)

The geologic history of Texas in the Tertiary is mainly one of building progressive thick layers of sand and mud southwestward into the continually downwarping Gulf of Mexico. This sedimentary wedge may be 50,000 feet thick under the Gulf Coast! These inclined wedges of sediment, arranged like a tipped stack of books, are traversed in order as you drive from Houston to San Antonio or Houston to Dallas. The outcrops are not very good, but the hilly Tertiary topography is noticeable in contrast to the flat Gulf Coastal plain.

Volcanoes spurted lava and ash in West Texas during the Eocene and Oligocene (24 - 58 million years ago), creating the Davis Mountains, Chisos Mountains, and many other peaks and ranges in the Big Bend Country. A later episode of structural relaxation and pull-apart created a series of block mountains and linear valleys, which account for much of the topography seen today in West Texas.

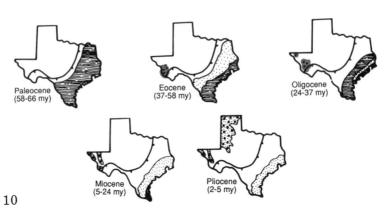

Paleocene (58-66 my)

Eocene (37-58 my)

Oligocene (24-37 my)

Miocene (5-24 my)

Pliocene (2-5 my)

A late Tertiary resurgence of uplift in the Rocky Mountains sent huge volumes of sediment eastward to blanket the Panhandle. The resultant Ogallala formation underlies the High Plains and is the main aquifer in this area.

The Tertiary period is the age of mammals, and all the common animals we know today arose in the Tertiary; primates, rodents, seals, elephants, horses, rhinos, pigs, hippos, camels, cattle, and deer. At the same time, the appearance of grasses in the Tertiary profoundly affected the development of many of these mammal groups. Near the close of the Tertiary, a spear-tossing, Ford-driving primate burst upon the landscape, some say to change the scene forever.

Subsequent stream erosion during high-moisture periods of the Ice Age (Pleistocene: 2 million - 10,000 years ago) dissected the High Plains and carved Palo Duro Canyon, the Caprock Edge, the breaks of the Canadian River, and entrenched the major streams across Texas.

QUATERNARY:
2 Million Years Ago to Today
(Pleistocene, Holocene)

Finally, as the Ice Age glaciers melted, sea level rose, inundating Gulf Coast river mouths to form bays and estuaries. Extensive barrier islands grew along the Texas coast as sand, brought to the sea by rivers, was spread along the new shoreline by the action of waves and longshore drift.

INTRODUCTION TO THE GEOLOGY OF TEXAS

The geologic panorama of Texas is as wide as the big state itself, sweeping from volcanic mesas and thrusted mountains in the west, to red canyons of the Panhandle, along tropical sand barriers of the Gulf Coast, and across central limestone plateaus onto hard granitic terrain in the center of the state. Rocks of all ages, from crystalline gneiss of ancient Precambrian time to the loose sand of modern beaches are found at the surface in the state, as well as every major rock type from igneous to metamorphic to sedimentary. Moreover, Texas is blessed with an array of natural geologic resources, ranging from its famous oil and gas fields to salt, sulfur, lignite, building stone, sandstone, gravel, sand, clay, uranium, other minerals, and water.

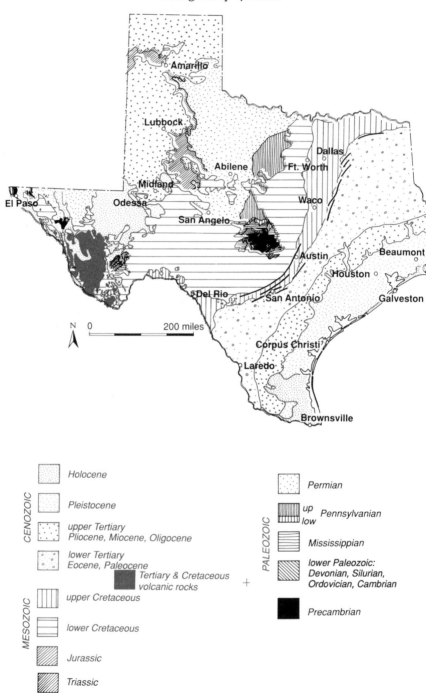

Geologic map of Texas.

CENOZOIC
- Holocene
- Pleistocene
- upper Tertiary
  Pliocene, Miocene, Oligocene
- lower Tertiary
  Eocene, Paleocene
- Tertiary & Cretaceous
  volcanic rocks

MESOZOIC
- upper Cretaceous
- lower Cretaceous
- Jurassic
- Triassic

PALEOZOIC
- Permian
- up
  low Pennsylvanian
- Mississippian
- lower Paleozoic:
  Devonian, Silurian,
  Ordovician, Cambrian
- Precambrian

+

*Line of buried Ouachita
Mountains across Texas.*

For Texans who live in large cities or for the out-of-staters who believe the stereotypical Texas of endless, featureless, windswept prairies covered by cattle and oil derricks, it may come as a surprise to learn of the state's topographic variety, geologic diversity and beckoning scenery.

For an overview of the geology of Texas, begin by looking at the geologic map of the state, which shows the ages of the main rocks at the surface. Several major patterns give clues about the principal aspects of Texas geology. First, notice the sweeping 'S' curve line through the center of the state, running from Dallas-Fort Worth, past Austin and San Antonio, and westward toward the Big Bend of the Rio Grande. Note how rocks east of the line roughly parallel it, are all younger than Cretaceous in age, and get progressively younger toward the Gulf of Mexico. The 'S' line marks the edge of the North American continent in Cretaceous time (60 to 100 million years ago) when the Gulf of

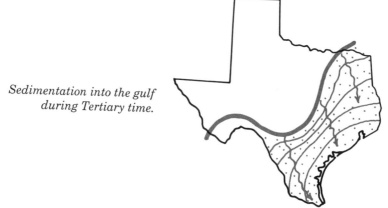

*Sedimentation into the gulf
during Tertiary time.*

13

Configuration of continents during the past 350 million years. Continental drift and sea floor spreading are results of thermal churning in the mantle beneath the earth's crust.

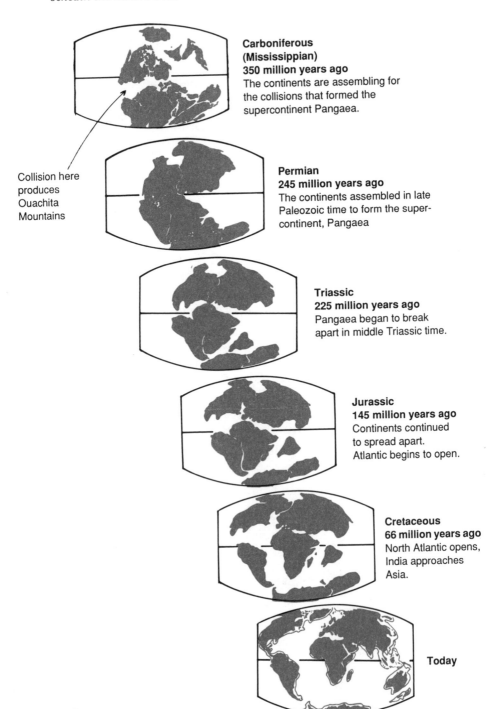

Carboniferous (Mississippian) 350 million years ago
The continents are assembling for the collisions that formed the supercontinent Pangaea.

Collision here produces Ouachita Mountains

Permian 245 million years ago
The continents assembled in late Paleozoic time to form the super-continent, Pangaea

Triassic 225 million years ago
Pangaea began to break apart in middle Triassic time.

Jurassic 145 million years ago
Continents continued to spread apart. Atlantic begins to open.

Cretaceous 66 million years ago
North Atlantic opens, India approaches Asia.

Today

Mexico formed in the pull-apart zone between North and South America, during a great worldwide episode of ocean-spreading, continental drift, and mountain building. This curved line also follows quite closely, and not fortuitously, a tightly folded mountain belt, which is buried across much of Texas, but extends from the Ouachita Mountains of southern Oklahoma, underground across Texas, to southwest Texas, where the range reappears around the town of Marathon, north of Big Bend National Park. The rocks east and south of buried Ouachita Mountains represent the great, continuous sedimentary filling of the Gulf of Mexico since 60 million years ago, as layer upon layer of gravel, sand, silt and clay were deposited into the Gulf by rivers carrying engorged sediment loads eroded from the high terrain of the newly uplifted Rocky Mountains. This wedge of sediment, estimated to be up to 40,000 feet thick, grew gulfward age by age throughout the Tertiary period (hence the age bands) and continues today as sediment is added to the Gulf by the Colorado, Brazos, Sabine, Pecos, Rio Grande, and other Texas rivers. The upper and lower Gulf Coast, coastal plain, and modern beach, barrier, lagoon, and delta systems are all the product of this immense sediment-wedge-building process.

In the organic-rich sediments of the deeply-buried Tertiary wedges, oil and gas were formed, while plant-rich sediments have yielded abundant lignite. A soft layer of salt, originally laid down during Jurassic time in the first shallow pan of the early Gulf of Mexico underlies the thick Tertiary rocks; in many places it penetrates upward through the sediment wedge with spike-like tentacles forming the famous Texas Salt Domes where oil and gas and sulfur have been yielded up for human use.

As the Ouachita Mountains rose 300 million years ago, the crust west of the mountains sank to form a series of deep basins in which thick sections of limestone accumulated along with sand and clay sediment wedges shed from adjacent ranges. The abundant oil and gas found in west and northwest Texas is a product of the deep burial and geologic cooking of these limestones and other beds of organic-rich sediments. The limestones are best seen in the Guadalupe and Delaware mountains of West Texas, while bright red sediments color canyons in the northwest and the north central part of the state.

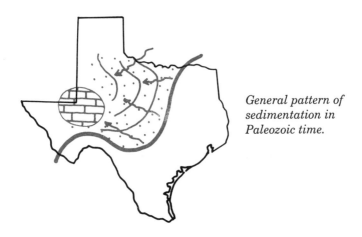

General pattern of sedimentation in Paleozoic time.

The Ouachita line has deeper roots, however, than just being the one-time edge of the Gulf of Mexico, for the line follows the trend of the mighty Ouachita Mountains which once crossed Texas, but is now worn down and buried beneath sedimentary rocks. And, the Ouachita Range itself follows a deep zone of weakness in the earth's crust, because this line has been the location for not one, but several episodes of continental collision and separation throughout the geologic past, beginning about a billion years ago in the distant Precambrian past.

Rocks uplifted during the Ouachita mountain building episode about 300 million years ago can be seen in the west around

The Appalachian-Ouachita-Marathon Mountain system.

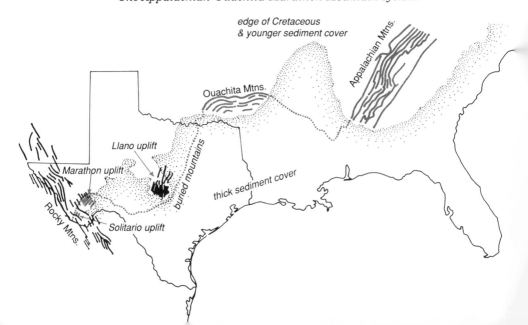

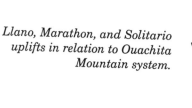

Llano, Marathon, and Solitario uplifts in relation to Ouachita Mountain system.

Marathon, Texas in the area known as the Marathon uplift. Similar rocks are seen in Central Texas west of Austin in the rocky terrain of the Llano uplift. But, also in the Llano area, old igneous and metamorphic rocks tell an even more ancient story of mountain building which occurred nearly one billion years ago!

For nearly 200 million years, quiet erosion wore the Ouachita Range nearly flat, and Cretaceous seas laid thousands of feet of limy rock over the old range. Then the earth heaved up one last time—about ten million years ago—to elevate the Cretaceous rocks nearly 2,000 feet above sea level, forming the Edwards Plateau. The fracture along which this last uplift occurred is known as the Balcones fault zone. It passes near San Antonio, Austin, and Waco and, not coincidentally, follows the trend of the Ouachita line.

North-South seaway connected the Gulf of Mexico with the Arctic Ocean at various times during Cretaceous time.

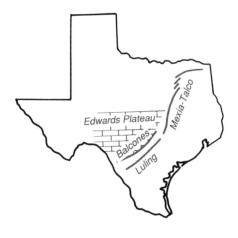

*Balcones, Mexia-Talco,
Luling fault zones, and
uplifted Edwards Plateau.*

While Cretaceous seas spread limestone, sandstone, marl, and mudstone over much of Texas, the crunch of the Rocky Mountains was being felt in West Texas, where long, linear mountain ranges pushed upward to form a series of northwest-southeast structures clearly seen on the geologic map. The uplift in these ranges pushed old rocks to the surface, and Precambrian-age sedimentary, metamorphic and igneous rocks can be seen in the Franklin Mountains near El Paso, the Sierra Diablo range north of Van Horn, and along Interstate 10 just west of Van Horn. Eastward, earthquake-induced fractures along the Balcones fault became avenues for the escape of lava from deep within the earth, and a series of subsea volcanic explosions rocked the Cretaceous seas, sending plumes of ash, rock, and steam upward. On the geologic map, the dots of these

*General trend of Rocky
Mountain structures in
West Texas.*

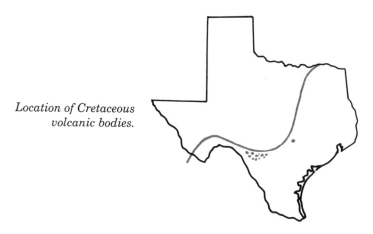

Location of Cretaceous volcanic bodies.

Cretaceous volcanoes stretch from Austin to Uvalde on the geologic map.

But the real volcanic episode for Texas was yet to come in the Eocene and Oligocene epochs (50 million years ago), when huge volumes of lava blasted to the surface along fractures in West Texas. Volcanoes, immense lava sheets, and thousands of lava domes now dominate the west Texas landscape from Big Bend National Park to Alpine, the Davis Mountains and northward to the New Mexico border.

Later in the Tertiary period, western North America underwent a period of relaxation, wherein linear basins and ranges were created along a series of normal faults. Block mountains and intervening valleys from this episode are common in West Texas.

Eocene - Oligocene volcanoes.

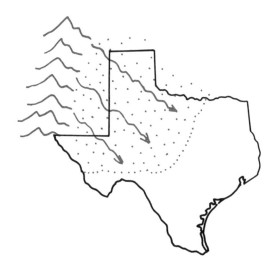

A sediment wedge built eastward across the Panhandle in the Tertiary from Rocky Mountain highlands to the west.

From the high Rocky Mountains in Colorado and New Mexico, rivers charged across the Panhandle of Texas laying down a vast apron of gravel and sand that extended nearly to the point where Dallas and Fort Worth now stand. This inclined "gangplank" of sediment covered the colorful older Permian and Pennsylvanian-age rocks, until the same regional uplift ten million years ago that elevated the Edwards Plateau, raised the Panhandle, and started a new cycle of intense erosion across Texas. For the last ten million years the Red River, Colorado, and Brazos rivers have been eroding back the edge of the Panhandle high plains, until today, the steep caprock escarp-

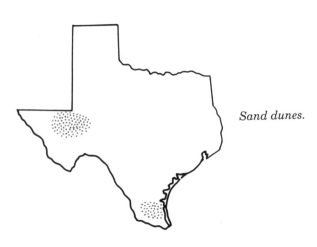

Sand dunes.

ment is about 200 miles west of its original edge near Dallas - Ft. Worth. The canyons along this escarpment form some of the most enchanting and colorful scenery in Texas; state parks, such as Palo Duro Canyon and Caprocks Canyon extoll this geologic beauty.

The tale of Texas geology is not complete, however, until we consider geologic processes operating today. The wind has moved and shaped sand into marvelous dunes and sand sheets between Corpus Christi and Brownsville. West of Midland and Odessa, Monahans State Park is a great place to study wind-blown deposits. A different kind of wind, the Hurricane, molds and shapes the Texas coast, where barrier islands, lagoons, bays, and sand dunes can be seen along a 1,000 mile stretch. Adding to the shape of the coastal environments are the additional forces of waves, tides, and river flow. Geologists have studied these modern Texas environments closely in order to understand exactly how surface forces interact with sediments. It is only through this understanding that ancient rocks, deposited in similar environments, can be recognized and interpreted.

OTHER EFFECTS OF GEOLOGY
How Geology Influences the Environment

The basic geologic patterns across the state have a strong bearing on other environmental elements such as weather, topography, and vegetation.

The most direct effect geology has is on the topography, and once you have seen the geologic map of the state, the adjacent topographic map makes much more sense. Highest elevations are in the west, where mountain ranges such as the Davis Mountains (8,300 feet), the Chisos Mountains (7,800 feet) in Big Bend National Park, the Guadalupe Mountains in Guadalupe Mountains National Park where Guadalupe Peak (8,751 feet) is the tallest in the state, and a host of other ranges, stand over 5,000 feet above sea level. The Panhandle Plains rise topographically toward their Rocky Mountain source, reaching elevations of 4,000 feet in the northwest corner of the Panhandle, whereas the Edwards Plateau in the center of the state is about 2,000 feet above sea level. In general, the topography is highest in the west and northwest, and decreases eastward to about 300 feet above sea level at Texarkana; it continues southeastward to the Gulf Coastal Plain which slopes gently toward sea level.

The observed topography is the product of two factors, rocks and erosion. Guadalupe Peak, for example, stands high because the rocks were uplifted, but equally important, because its reef limestones are hard, and resist erosion better than softer surrounding rocks. The topography of the uplifted, north-south ranges in West Texas stands out clearly because of the faulted structure of these ranges, but again, they contain hard rocks which differentially resist erosion. It is also easy to spot the distinctive dome-like topography of the volcanic intrusions

Physiography of Texas.

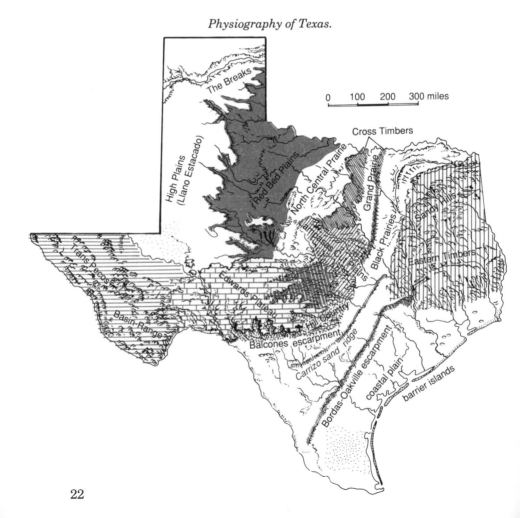

Average Annual Rainfall
(inches per year)

around Big Bend National Park. The hard basalt bodies started out roughly dome-shaped, and erosion has had a tougher time attacking these than the less resistant shales surrounding them. This interaction of fundamental geologic structure with erosion to create topography is repeated many times across Texas. Keep these two factors in mind as you interpret the landscape from your car window.

Weather fronts sweep across Texas mainly from west to east, though splashes of water-laden clouds frequently move landward from the Gulf. Hurricanes encounter the coast less frequently, but as is described in more detail in later chapters, they have a significant impact on the coastal topography. The precipitation pattern is quite uniform, as shown on the map. Annual rainfall increases west to east, from eight inches per year in the dry western country around El Paso to over 50 inches per year in the eastern semi-tropical forests near Beaumont. The lines of equal rainfall track almost directly north-south, or roughly perpendicular to the prevailing west to east weather fronts. However, notice the circular, higher rainfall anomaly in West Texas, centered over the Davis Mountains. Geology does affect the weather! The high Davis Mountains force the westerly winds to rise to climb over the peaks, and in the process the flowing air is cooled, and its contained moisture is forced out, dropping as rain on the flanks of the peaks. A detailed rainfall map would show the same increase in moisture on most of the ranges in West Texas and the increased moisture, hence denser vegetation is particularly noticeable on the beautiful Chisos Mountains in Big Bend National Park, and the craggy peaks of Guadalupe National Park. The geology controls the topography, which controls the local weather, which controls the local plants and animals.

One more environmental connection is worth making here. A close look at the vegetation map of Texas, reveals familiar patterns that closely follow the patterns found on the geologic map of Texas! Annual temperature and rainfall are dominant influences on the distribution of vegetation, but underlying soils, which are strongly influenced by the rocks, also control

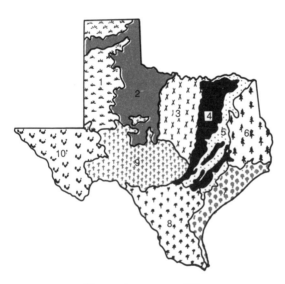

Vegetation areas of Texas.

Vegetation Areas

1. High Plains - grass, sagebrush, mesquite, and yucca
2. Rolling Plains - mesquite, woodland and grass prairies
3. Cross Timbers and Grand Prairie - post and live oak, woodlands, and grass prairie
4. Blackland Prairie - grassy plains, oak, pecan, elm, streamside woodlands
5. Post Oak Belt - oak, elm, pecan, walnut woodlands, and mainly grass prairies to the south
6. Piney Woods - thick pine stands and oak, hickory, magnolia, gum, and hardwood forests
7. Gulf Prairies and Marshes - salt grasses, oak and elm forests
8. South Texas Brush Country - mesquite, cactus, small oaks, and bunchgrass
9. Edwards Plateau and Hill Country - highly variable vegetation: oak, cedar, mesquite woodlands, grass prairies, and cypress waterways
10. Desert Mountains and Basins - arid lechuguilla, ocotillo, yucca, mountain pines, ponderosa pines, and junipers

24

plant distribution. For example, the broad sweep of eastern pine forests follows the sandy ridges of Tertiary rocks in the area, because pines like well-drained sandy soils. The thin soils on top of broad expanses of rocky limestone terrain in the Edwards Plateau of central Texas support stands of live oaks, which prefer calcareous soils. The deciduous Post oaks, on the other hand, prefer sandier soils, whereas mesquite grows preferentially on clay-rich soils where many other trees have a difficult time competing. Rich alluvial soils parallel to major rivers commonly host more lush vegetation than nearby hillsides.

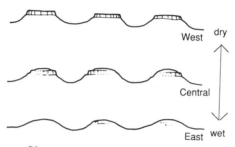

Change in hill form across Texas.

But, finally, the cycle is complete, because the geology itself is directly affected by rainfall and vegetation which control the intensity and style of erosion that attacks rocks! In the dry west, where rainfall and plants are sparse, mechanical erosion by the wearing action of intense run-off after cloudbursts is the main erosive force. In the east where rainfall is high, abundant plants create thick humic soils which hold back water and prevent intense runoff. Water held in the ground, however, causes more intense chemical erosion on rocks, breaking them into small bits (soil). As a result, west Texas rocks are mechanically worn down, and hillslopes take on the form of mesas. Large rocks simply break and fall downslope under gravity. In east Texas, chemical erosion causes 'softer', rounded hills, because rocks disintegrate to small particles which move downslope via slow creep.

As you drive across Texas, you can actually see the difference in hill forms from one side of the state to the other.

WHY SO MUCH OIL IN TEXAS

Texas, blessed with an enormous national resource in oil, straddles a buried mountain range whose existence explains, in large part, the occurrence of the oil.

About 300 million years ago, (Pennsylvanian period), the northwest half of Texas was part of a continent that collided into another continental piece, via the process of plate tectonics, to form part of the supercontinent Pangaea. In the process a mountain range was heaved upward along the collision line. This range we now call the Ouachita Mountains. Pieces of this range are still exposed at the surface in Oklahoma, in the Llano uplift northwest of Austin, and around Marathon in southwest Texas. In between, the range lies buried in Texas beneath piles of younger sedimentary rocks.

On the continent side of the uplifted Ouachita Range, in west and northwest Texas, the crust was downwarped in several places in compensation for the adjacent uplift. These downwarps,

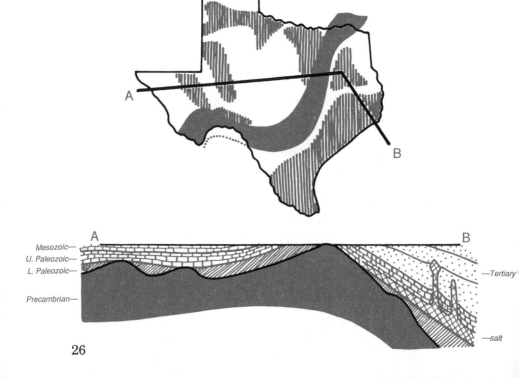

Sedimentary basins of Texas.

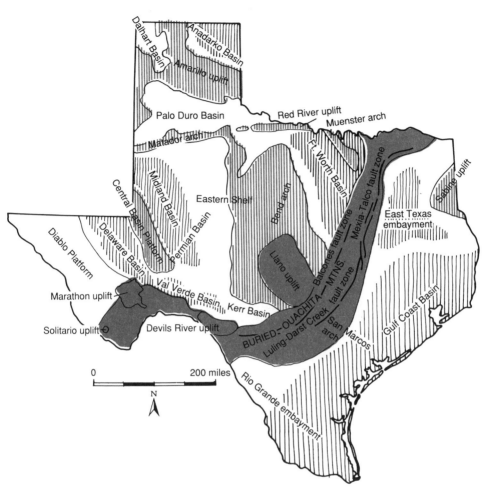

Structural geology of Texas. (Most of these are subsurface features known by oil well drilling.)

or basins, continued to settle over several millions of years, receiving organic-rich deposits of calcareous mud from marine organisms living in shallow seas that inundated these low spots. Reefs and banks of limestones formed around the edges of the basins from the growth of abundant reef-building animals. The organic-rich mudstones, and even the basin-edge limestones, were the source for much of the oil in West Texas, while the cavernous limestone reefs and banks became the reservoirs to store the oil. A magnificent example of these rocks and their story is to be found at Guadalupe National Park in West Texas.

By 200 million years ago (Jurassic period), Pangaea began to pull apart in the great continent-wrenching episode that generated the Earth's present configuration of continents and ocean basins. The Gulf of Mexico began to drop away from the old Ouachita Mountain line as North America and South America separated. At this early stage, the Gulf was only narrowly connected to the other oceans, and upon this pan of sometimes sea and sometimes dry flat, thick layers of salt were evaporated. Thus formed the Jurassic-age Louann salt.

The Gulf of Mexico continued to deepen and thousands upon thousands of feet of sediment poured into the basin from the emergent North American continent. These were mainly organic-rich muds, similar in kind to the delta deposits at the front of today's Mississippi River. River and shoreline sands were also laid down in this process to form the source-reservoir combination that contributed to east Texas' gigantic oil fields. But, the thick pile of Mesozoic and Cenozoic-age sedimentary deposits bore down heavily on the Louann salt through time, and it responded by heaving upward in tall spires, banks, and

Major oil and gas reservoirs in Texas.

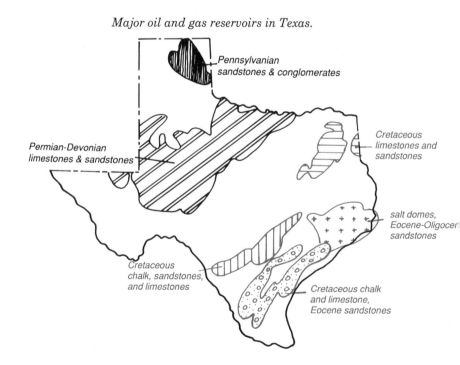

Pennsylvanian sandstones & conglomerates

Permian-Devonian limestones & sandstones

Cretaceous limestones and sandstones

salt domes, Eocene-Oligocer sandstones

Cretaceous chalk, sandstones, and limestones

Cretaceous chalk and limestone, Eocene sandstones

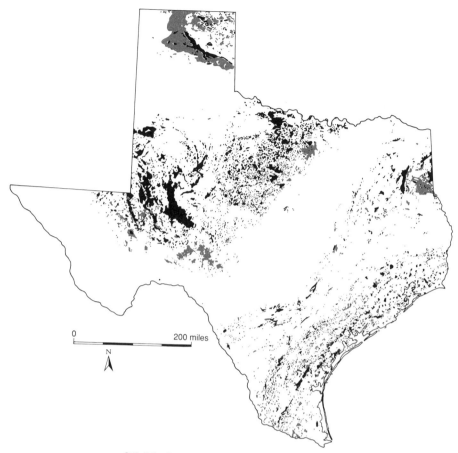

Oil (black) and gas (red) fields of Texas.

domes, which cut through the overlying sediments to form salt domes. And, around these domes, oil was trapped in profusion.

Hence, Texas has a great amount of oil and gas because there has been not one, but two great periods of hydrocarbon generation. First, the oil in West Texas was generated and trapped in a number of basins which developed in Paleozoic time in response to Ouachita mountain building. The second great period of oil generation is the product of later, Mesozoic tectonic forces, which opened the Gulf of Mexico and allowed the deposition of thick organic-rich sediments into the Gulf.

The history of oil use in Texas goes back quite a ways, for the native American Indians living here found and used oil from surface seeps long before Europeans arrived. Probably the

Texas crude oil production peaked in 1972 at 3.72 million barrels per day.

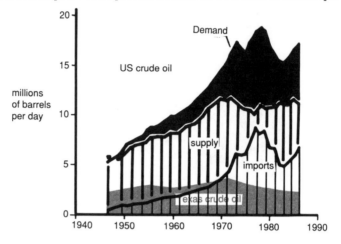

Four geologic elements are needed to form an oil or gas field.

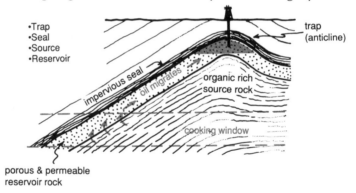

Types of oil traps.

anticline

fault

truncation

pinch out

salt dome

earliest record of Europeans' use of oil is by the survivors of Spanish explorer Hernando De Soto's expedition, who caulked their boats with tarry oil in 1543, near Sabine Pass in the southeast corner of the state.

The first well to produce oil was drilled in 1866 by Lynn T. Barret near Melrose in Nacogdoches County in east Texas. In 1867, Amory Starr and Peyton F. Edwards brought in a well at Oil Springs in the same area, giving Nacogdoches County the honor of the first commercial oil field, production, refinery, and pipeline activity in the state.

However, the first major oil discovery came in 1894 when the city of Corsicana, south of Dallas—Ft. Worth, tried to drill a water well and discovered the Corsicana oil field instead!

In 1901, the first great Texas gusher and giant field was brought in by Captain Anthony F. Lucas, who drilled Spindletop near Beaumont. In 1900, Texas produced a mere 836,000 barrels of oil, and in 1901, 4,400,000 barrels. By 1902, Spindletop poured out 17.5 million barrels of oil which accounted for 94% of Texas' total production that year.

A gas wellhead, or "Christmas Tree," as it is commonly called.

The biggest oil field in Texas, the East Texas Field, was discovered near Kilgore, in Rusk County, in 1930 by the veteran wildcatter, C. M. (Dad) Joiner. The East Texas Field still produces oil today, but has been surpassed in daily output by the Yates field, also a giant field, discovered in West Texas in 1926.

Texas has 52 giant fields, each having an estimated ultimate recovery of 100 million barrels of oil or more. The state produced over 800 million barrels of oil and more than 5 trillion cubic feet of natural gas in 1986, though production of both gas and oil peaked in 1972, and the annual yield has been decreasing ever since.

(Data from Texas Almanac, Dallas Post, 1989-1990)

Energy resources in Texas (excluding oil and gas).

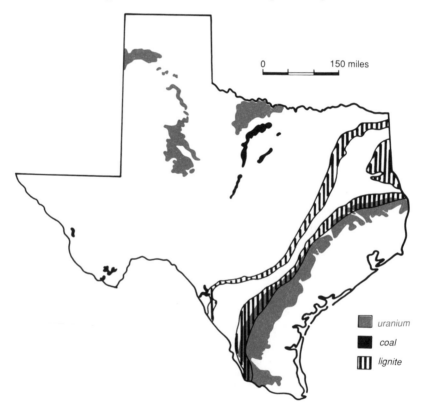

0   150 miles

■ uranium
■ coal
Ⅲ lignite

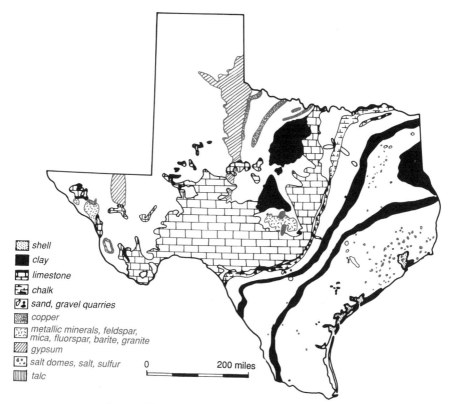

shell
clay
limestone
chalk
sand, gravel quarries
copper
metallic minerals, feldspar, mica, fluorspar, barite, granite
gypsum
salt domes, salt, sulfur
talc

0   200 miles

Areas of known or potential mineral reserves.

Mineral Resources of Texas

Aluminum	Dolomite	Marble	Serpentinite
Asbestos	Feldspar	Mercury	Shell
Asphalt	Fluorspar	Mica	Silver
Barite	Gold	Molybdenum	Sodium Sulfate
Basalt	Graphite	Peat	Strontium
Beryllium	Grinding Pebbles	Perlite	Sulfur
Brine	Gypsum	Phosphate	Talc & Soapstone
Building Stone	Helium	Potash	Tin
Cement Materials	Iron	Pumicite	Titanium
Chromium	Lead	(Volcanic Ash)	Tungsten
Clays	Lignite	Salt	Uranium
Coal	Limestone	Sand, industrial	Vermiculite
Copper	Magnesite	Sand and Gravel	Zeolites
Crushed Stone	Magnesium	(Construction)	Zinc
Diatomite	Manganese	Sandstone	

•Some mineral resources are actively produced in large quantities today, such as building stone, sulfur, shell, sandstone, sand, lignite, gypsum, brine, clays, crushed stone, salt and limestone. Others are minor, or are potential resources. A few, such as mercury and iron, have been produced in past times, or were produced in greater quantities than today.

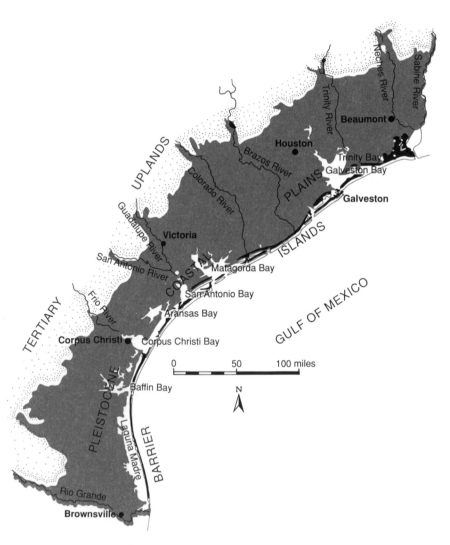

Geologic map of the Texas Gulf Coast.

II
The Gulf Coast
Salt Domes, Rivers, Beaches, Dunes, and Barrier Islands

The Texas Gulf Coast is a strip of low, flat terrain more than 300 miles long by 50 to 100 miles wide, lying adjacent to the shoreline of the Gulf of Mexico. The surface is indeed flat, but not quite horizontal, as the coastal plain tips gently gulfward at about five feet or less per mile. Despite the fact that few rocks are seen along highways crossing the coastal plain, this mud and sand flat has a marvelous geologic story to tell.

The coastal plain is the mere surface expression of a huge wedge of sediment dumped into the Gulf of Mexico by Texas rivers for the past few million years. Moreover, the process of coastal building in Texas has been going on in earnest for at least 65 million years, extending the land of North America over 250 miles into the Gulf of Mexico since that time. Continuous mountain erosion, river transport of sediment, and deposition in the deep hole of the Gulf, is the earth's way of saying

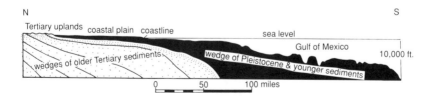

Cross section of the Gulf Coast.

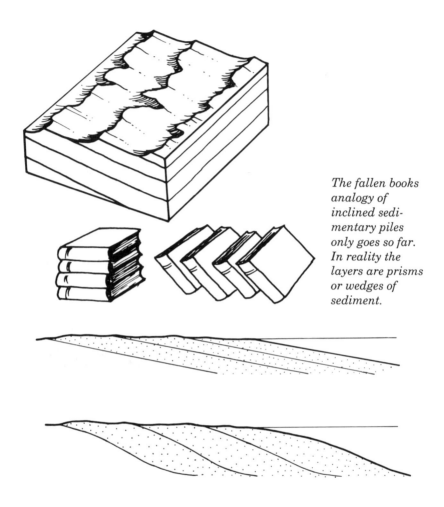

The fallen books analogy of inclined sedimentary piles only goes so far. In reality the layers are prisms or wedges of sediment.

it doesn't like topography! Over geologic time, high mountains are reduced to flat plains, while deep oceans are filled up until they too are flat plains.

The coastal plain is the latest expression, then, of a series of sloping sediment wedges that have built gulfward since Cretaceous time (65 million years ago). However, the recent geologic history of the coastal plain is closely tied to the rise and fall of sea level. For the last few hundred thousand years, sea level has risen and fallen in response to the growth and melting of the great Ice Sheets that once nearly covered North America. As ice melted, water was added to the oceans, causing sea level to rise. Conversely, as ice sheets grew, rainwater was tied up in ice, less water returned to the oceans via rivers, and sea level dropped.

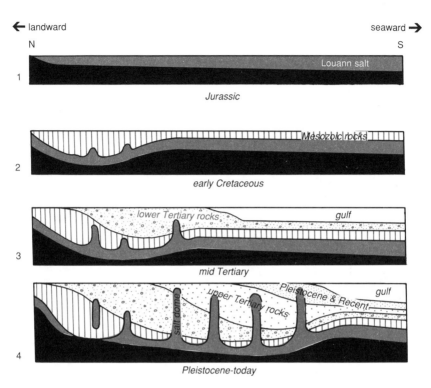

N   S

1

Louann salt

Jurassic

2

Mesozoic rocks

early Cretaceous

3

lower Tertiary rocks   gulf

mid Tertiary

4

salt dome   upper Tertiary rocks   Pleistocene & Recent   gulf

Pleistocene-today

History of development of Gulf of Mexico and sediment wedges.

About 135,000 years ago, during a warm (warmer than today) interglacial period, sea level was higher by about 25 feet than it is today, and the coastline was then 20 miles inland from today's coastal position. Coastal plain rivers built channels and floodplains at higher elevations to match the elevated coastline. Then sea level dropped as the last Ice Age reached its frigid apex about 18,000 years ago. The lowered coastline moved seaward many miles and rivers had to adjust their gradients accordingly. They did so by cutting downward to match the lower coastline level, and the coastal plain became a wider surface deeply dissected by entrenched rivers. At 18,000 years ago the great northern hemisphere ice sheets began to melt, and sea level rose since that time, reaching present levels only about five thousand years ago. The rising sea drowned river mouths and therein created bays and lagoons along the coastline, while rivers adjusted to the higher sea level by partly filling in their deep valleys. Present sea level is not as high as it was 135,000 years ago, because today's climate is not as warm as it was then. Consequently, a large volume of the

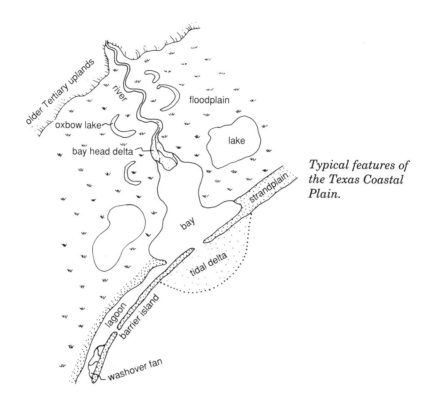

Typical features of the Texas Coastal Plain.

earth's water is still entrapped in glacial ice, mainly in Greenland and Antarctica, and if this remaining ice were to melt, the Texas coastal plain would be covered once again by shallow sea water.

MEANDERING RIVERS

Like an undulating snake or loose garden hose, rivers crossing the Texas coastal plain twist and turn in a sinuous pattern. Rivers tend to meander in this fashion because their banks aren't perfectly uniform and the water flow is turbulent or highly irregular. Even if the river channel started out perfectly straight, it would soon develop bends and meander, because any small irregularity in the channel would deflect the stream flow. As soon as this happens, one bank is pounded harder by the force of the water, causing greater bank erosion on that side, which puts a slight bend in the channel. Once started, the current hits the outer bend with greater force than the inside bend, and over time, the bend grows larger and becomes more accentuated. As the bank on the outer bend erodes sideways, the

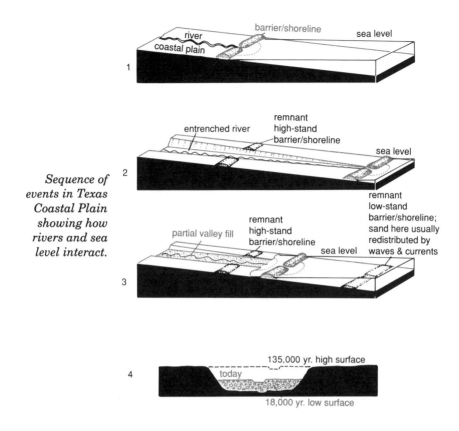

Labels in figure:

1. coastal plain, river, barrier/shoreline, sea level

2. entrenched river, remnant high-stand barrier/shoreline, sea level, remnant low-stand barrier/shoreline; sand here usually redistributed by waves & currents

3. partial valley fill, remnant high-stand barrier/shoreline, sea level

4. 135,000 yr. high surface, today, 18,000 yr. low surface

sediment carved from one bend is deposited in the slow-moving water on the next downstream inside bend. This sandy pile is called a point bar. Erosion of the outer bend and deposition on the inner bend cause the channel to migrate sideways through time. The motion of a loose garden hose nicely simulates the motion, through time, of a meandering river.

As meander loops grow, they become nearly circular in plan, and eventually the river cuts through the narrow neck to follow a straighter course, leaving the old loop to form a C-shaped lake called an oxbow lake.

The mechanics of river channels are only one part of the floodplain story, however. Much sediment, mainly mud, is carried by the river and deposited on the adjacent plain during floods. The floodplain builds upward and seaward by the regular addition of these mud layers. The process is so inexorable that Texas Gulf Coast floodplain construction has added about 250 miles of land to the United States in the last 60 million years.

39

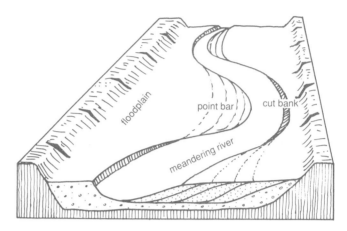

Anatomy of a meandering river.

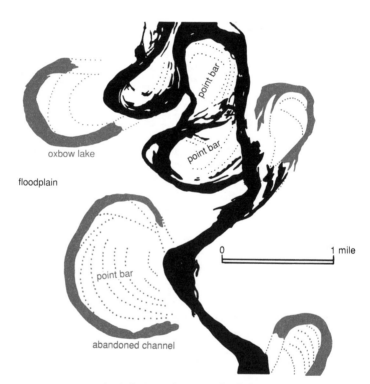

Aerial view of a meander belt.

How meander channels develop.

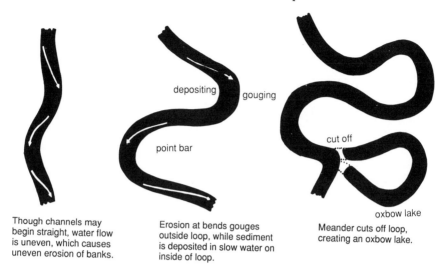

Though channels may begin straight, water flow is uneven, which causes uneven erosion of banks.

Erosion at bends gouges outside loop, while sediment is deposited in slow water on inside of loop.

Meander cuts off loop, creating an oxbow lake.

Roads traversing the Texas coastal plain cross a number of meandering rivers. Good views of point bars, channels, and steep outer bends can generally be seen from the highway bridges over the rivers.

GEOLOGIC HISTORY OF THE TEXAS BARRIER ISLANDS

The Texas coastline is a beautiful, long, curved, sandy rim on the edge of the Gulf of Mexico. The rim itself is a string of elongated sandy beads, called barrier islands, that separate the mainland coast from Gulf waters. Padre Island, the longest single barrier island in the United States, stretches 113 miles under sunny Texas skies, where nearly a million visitors a year come to enjoy the sun, beaches, and natural beauty. The coastline and the adjacent barrier islands are a complex of ever-changing environments driven by a dynamic interplay of wind, water, tides, waves, streams, and storms. The coastline is not static. It moves and changes, growing here, retreating there, shifting one way, then another.

The great age of most geologic features is so often emphasized, that it probably comes as a surprise to learn the barrier islands are only a few thousand years old, actually about 4,500 - 5,000 years, based on radiocarbon dating of shells. How the barrier islands came into being

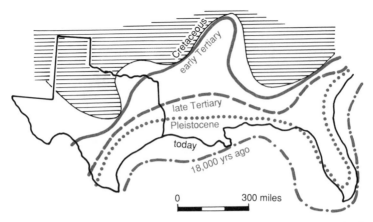

Map showing how the margin of North America has built progressively gulfward since Cretaceous time. The 18,000 year line shows where shoreline was when glacial ice was at a maximum and sea level was lower than it is today.

is a fascinating story, one intimately tied to far-away Ice Age glaciers and related changes in sea level.

About 18,000 years ago, sea level was about 300 to 450 feet lower than it is today, because a large amount of the earth's water was frozen in the great ice sheets which covered vast areas of North America and Europe. And, because of the lower sea level, the Texas shoreline lay about 50 miles farther gulfward than its present position. Rivers flowed across the exposed shelf area, while upstream they incised deeply into the coastal plain in order to flow with a smooth gradient to the lower and more distant shoreline.

As world climate warmed between 10,000 and 5,000 years ago, the ice melted, water was returned to the oceans, sea level rose to within 15 feet of present level. As the sea migrated landward to again cover the shelf, river mouths were drowned to form today's bays and estuaries. Upstream, rivers dumped their sediment load in the deeply carved valleys to adjust their gradients to meet higher sea level and to accommodate the shorter path to the coast.

After sea level rose, abundant sand lay on the submerged shelf, where it once was deposited in river channels, at river mouths, and along old shorelines. Now drowned and subjected to waves and storms, the sand was pushed and eroded and reworked toward shore. Sand built up, chiefly on the high submerged drainage divides between the old Ice Age rivers. These short, incipient barrier island segments continued to grow and merge to form the long barrier islands of today's coast.

Geologic history of Texas Barrier Islands.

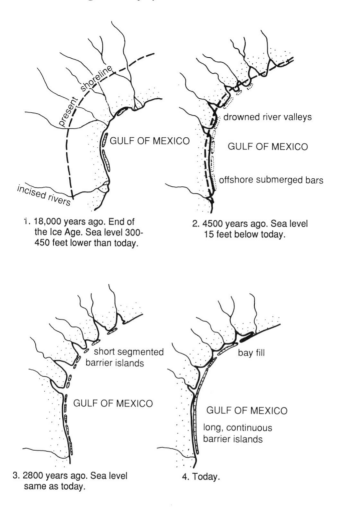

1. 18,000 years ago. End of the Ice Age. Sea level 300-450 feet lower than today.

2. 4500 years ago. Sea level 15 feet below today.

3. 2800 years ago. Sea level same as today.

4. Today.

The actual process of barrier island construction has been debated among geologists for many years, but three main ideas emerge as most plausible. First, barrier islands may form from submerged offshore sand bars. On some coasts, as many as three ridges of offshore bars are found. These could migrate shoreward with the addition of sand to eventually emerge as barrier islands. Second, barriers could form by lateral accretion from headlands as sand is moved longshore (longshore

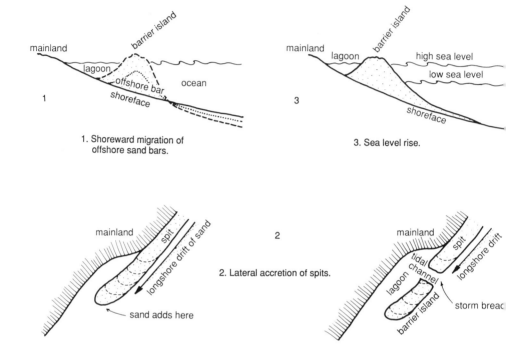

Ideas about the formation of barrier islands.

drift) by the consistent oblique approach of waves which strike the shoreline. Third, barriers may be developed by drowning the area on the landward side of mainland beach sand ridges, as sea level rose.

All three processes operate on shorelines today, and a combination of them is the likely explanation for the development of the long barrier islands of the Texas coast.

LONGSHORE DRIFT

As you stand on the beach and feel the energy of the waves breaking along the shore, it is readily apparent that sand trapped in this dynamic circulation is in constant motion. Rivers bring their sandy offerings to the sea, and in response, ocean waves return the sand to the shore. If waves always came straight in to the coast, sand would merely pile up on the beach, and the beach would grow uniformly

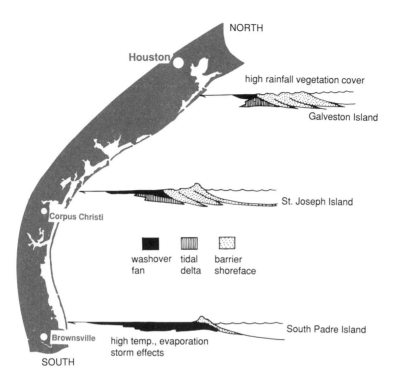

Change in barrier island sediments from north to south along the Texas coast.

seaward. But waves more commonly attack the shore at an angle, which moves sand along the shore. Think about one sand grain as it is propelled up the beach at an angle by an oblique wave. As the swash of water then falls back down the beach, the sand grain falls back, not at an angle, but straight down the beach towards the water. The grain has effectively been moved some distance down the beach by this arc of motion. Now multiply the one grain and the one wave by billions, and you can see how the whole beach is moving "alongshore." It is easy to see this process as you watch oblique waves strike any beach.

The long curve of the Texas barrier islands has a unique effect on longshore sand drift. Waves approach the Texas coast from different angles, of course, depending on the seasons and wind patterns, but the net wave approach is from the east-southeast. At the northern end of the coast, the net wave approach drives sand alongshore in a southwest direction, but because of the curve of the shoreline, sand at the southern end of the Texas coast is driven northward by the same wave approach. There is, therefore, a convergence of longshore drifting sand near the center of the barrier island arc. This zone of convergence is located at the central part of Padre Island. Another piece of evidence supports the convergence idea: sandy beaches change gradually to shell beaches near central Padre Island. Shells on Little Shell Beach (small shells) come from the north, while the large shells on Big Shell Beach (just south of Little Shell Beach) come from the south. The converging longshore drift thus tends to pile both shells and sand toward the center of Padre Island. A second process concentrates the shells—as the wind blows sand off the beach into the dunes, the shells are left behind as a "deflation lag."

Waves approaching the shoreline at an angle drive sand up the beach at an angle. But, the return flow of water and sand is straight down the beach. This arc of motion drives sand 'long-the-shore'.

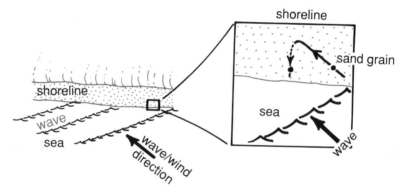

Convergence of longshore drift along the Texas coast.

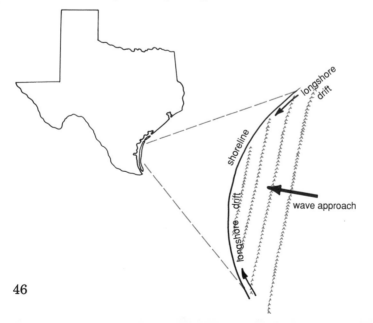

HURRICANES ALONG THE TEXAS COAST

Hurricanes are such major forces shaping the Texas coastline that the landscape of the Texas Coast cannot be understood without considering the effect of hurricanes.

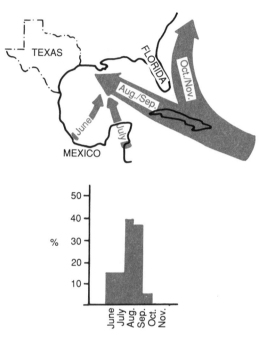

Early season hurricanes are rare, and tend to move north. The more frequent late season hurricanes generally come in from the southeast.

Since 1899 a significant hurricane has struck the Texas coast, on the average, a little more than once every other year. In the 30s and 40s the frequency of Texas hurricanes was much higher than average. Interestingly, these two decades were the warmest in North America in the 20th century, and because hurricanes derive their energy from the ocean surface, warmer temperatures (more energy) would be expected to increase the likelihood of hurricanes.

June through September is peak hurricane season in the Gulf of Mexico. Early summer hurricanes are more common along the Texas coast than elsewhere along the Gulf, due to hurricane origination in the warmer waters of the Caribbean at that time and a tendency for early season Atlantic hurricanes to follow a straight westward course. As the summer progresses, paths of Atlantic hurricanes tend to move eastward and northward along the United State's eastern seaboard.

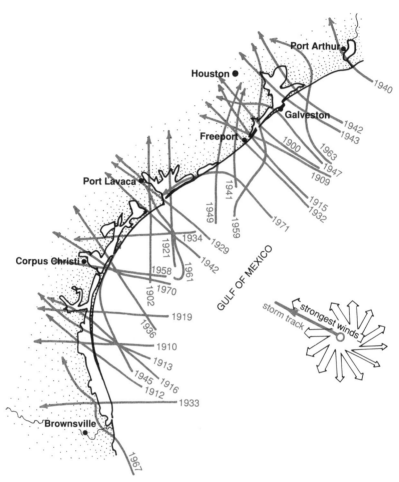

Tracks of hurricanes along the Texas coast from 1900 to 1979.

Hurricanes are great circular, counterclockwise motions of air that derive their energy from air rising off warmer (greater than 80° F) ocean waters, attaining wind speeds of 75 miles per hour and greater. The wind in hurricanes is asymmetrical, however, being most intense to the right of the center or "eye" of the storm track, because the effect of the counterclockwise circulatory winds is added to the forward motion of the storm system. Hence, the peak surge is always a few tens of miles to the right of the point of landfall of the storm. For example, Houston area residents will recall vividly the 10-15 inch rain that fell, dramatically flooding the east and northeast side of town (Spring area) during tropical storm Allison in 1989, while only 3-4 inches fell west of the city. The eye of that storm passed along the west side of Houston.

High winds in hurricanes are a major force on buildings, roofs, windows, etc. but water surge is probably the greater geologic force affecting the shoreline. As a hurricane moves across the ocean, the surface water is elevated because of the reduced atmospheric pressure. Wind stress also pushes water, creating a surge. Storm waves ahead of hurricanes can easily raise water level three feet before the storm arrives onshore. The elevated water level and waves at the shoreline do the most geologic work.

The environmental effects (property damage, flooding, shoreline erosion, etc.) are functions of wind intensity, rainfall, wave heights, and storm tide. The magnitudes of these effects are, in turn, dependent on the atmospheric pressure in the storm, its speed of travel across the ocean, angle of storm track to the coastline, duration of the storm, radius of the central funnel, shoaling effects of the continental shelf offshore and the height of the tide at the time of the storm.

Geologically, hurricanes affect the deposits, erosion, and form of the coastline. The flat profile of Galveston Island, Bolivar Peninsula, and Louisiana's adjacent Grand Isle, the most hurricane prone areas along the Gulf Coast, are the direct result of hurricane hammering.

Hurricanes work on the geology of coastal areas by flattening dunes and beaches, transporting and depositing sediment in coastal bays, opening channels through barrier islands, ripping up biological communities both on and offshore, and laying down resultant shell hash layers. They deposit washover fans in lagoons, and cause extensive inland, bay-head, and river flooding that, in turn, erodes and moves sediments.

Notice the difference in profile, presence of dunes and dune heights in driving from the low, mostly dune-free barrier islands and coastline at Galveston southwestward to the Padre Island area where steeper beach profile and high dunes are found. This change from northeast to southwest is attributable to two factors: more hurricanes to the northeast, and drier climate in the southwest, where lack of vegetation allows more sand to blow around freely.

SALT DOMES

Salt domes do not commonly reach the surface and therefore are not prominent features in the Texas landscape, but they do form an important part of the subsurface geology of the state.

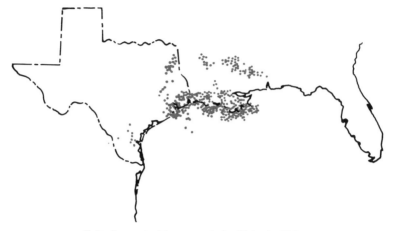

Salt domes in Texas and the U.S. Gulf Coast.

In what is now east Texas, a vast shallow sea lay to the east of the old Ouachita Range during upper Triassic time (about 200 million years ago). Because of the hot, dry climate then, the waters continually evaporated, leaving behind salty deposits of evaporite minerals, such as salt, gypsum, and anhydrite. Over time, thousands of feet of these salt deposits accumulated (the Louann salt), later to be buried beneath even greater thicknesses of younger sediments. Some geologists estimate the main layer of Louann salt may be 30,000 - 50,000 feet deep beneath the surface of the Texas Gulf Coast.

Salt has curious properties. We normally think of it in terms of hard, brittle little cubes, but when salt is subjected to heat and pressure, it becomes soft and plastic and can even creep and flow.

Salt dome basins in the Gulf of Mexico.

50

In the Gulf of Mexico, as the pile of younger sediment thickened over the Jurassic salt layer, the salt was buried deeper and deeper, and became hotter and softer and more pliable. The weight of the overlying sediment pile bore down on the salt, working much like your hand does when you push it down on a layer of wet mud. The salt oozed upward through the pressing sediment like the mud oozes up between your fingers, except the upward flowing salt created pillars, domes, ridges and needles, some rising upward to stand tens of thousands of feet above the mother lode.

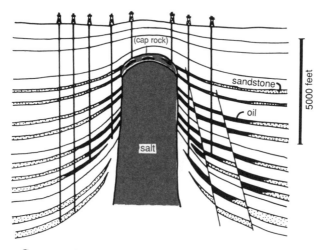

Cross section of Spindletop salt dome in East Texas.

For many years, Gulf Coast geologists have studied the mechanics of salt dome formation. Several dominant processes emerge: one is "diapirism", wherein salt domes penetrate upward through the thick layers of overlying sediment. The driving force is the weight of the sediments pushing down on the mother lode salt layer. The mobile, plastic salt squeezes upward like toothpaste, penetrating the sediment layers above to create salt domes, ridges, and pinnacles.

Modern Gulf Coast seismic records, shot over the continental shelf and slope of offshore Texas, have confirmed a second major process of salt dome construction known as "downbuilding". In this process, the overlying sediment continually collapses downward along arc-shaped faults to displace salt downward. Small sediment-thickened basins are formed, and salt domes and ridges are left standing high next to these basins, while displaced salt below the basins moves laterally downdip toward the toe-end of the continental slope.

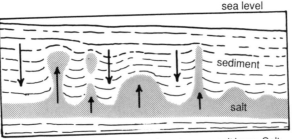

Diapirism — sediment load pushes down on salt layer. Salt squeezes upward, penetrating overlying sediment column.

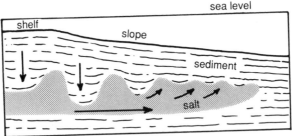

Downbuilding — sediment load pushes down on salt layer. Salt ridges and domes are left standing high next to small faulted basins which grow downward into salt. Displaced salt moves laterally basinward.

When the top of a salt dome is within a few thousand feet of the surface, ground water leaches cavities in the salt mass, and precipitates residues such as gypsum, calcite, anhydrite and sulfur. These minerals form a "cap rock" over the top of the salt dome.

Salt, sulfur, and gypsum are mined from the cap rock of many east Texas salt domes. Salt dome caverns are also used as natural underground storage sites for oil, radioactive wastes, and even valuable files and records.

Rocks at the top and along the flanks of salt domes have yielded billions of barrels of crude oil in east Texas. Spindletop, one of the earliest, largest, and most famous oil fields in Texas was discovered in 1901. The oil came from cap rock reservoirs of a large salt dome near Beaumont, Texas. It took 24 more years before oil was discovered in the sedimentary rocks which pinch out against the flanks of salt domes, setting off a new wave of exploration.

Salt domes were the object of the first "remote sensing" methods applied in oil exploration. Because salt is less dense than other sedimentary rocks, salt domes could be located by carefully measuring gravity variations and drilling where the gravity measured "low." Many salt dome oil fields were found by this technique.

Though surface expressions of salt domes are not common in east Texas where over 500 salt domes have been mapped in the subsurface, two exceptional mounds stand up clearly above the surrounding plains and can be visited easily by car. Damon mound rises 80 feet above the flat coastal plain west of Houston near the small town of Damon, located on U.S. 36 between Rosenberg and West Columbia. High Island is a caprock mound located east of Galveston Bay at the coastline near the south end of Highway 124. Four other surface mounds created by salt domes are located north of Houston, near Palestine and Marquez, a few miles from Interstate 45, but they are not as obvious as the mounds at Damon or High Island.

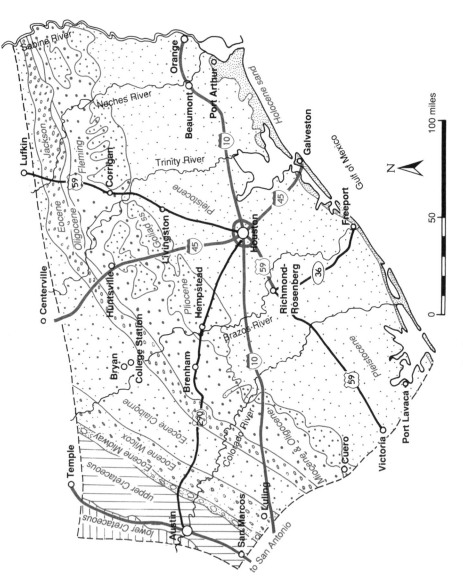

Geologic map of southeast Texas.

Southeast Texas
Upper Gulf Coast

Interstate 10
Houston — San Antonio
186 miles

Depending on where you start from in either San Antonio or Houston, the drive along Interstate 10 between the two cities is a bit less than 200 miles. The eastern half, between Houston and Columbus, where I-10 crosses the Colorado River, traverses the distinctly flat terrain of the Gulf coastal plain. You won't see any rocks along this stretch of highway, but to a geologist's way of thinking, the flatness of the coastal plain is as interesting a phenomenon as rugged mountains. The plain tells a story—one of eons of streams and meandering rivers carrying infinite loads of sediment from the Rockies and the interior of the North American continent, and depositing the sediment in swamps and low areas between rivers to form the flat coastal plain. The North American continent has been building continuously southward into the Gulf of Mexico in this manner for at least 60 million years; the coastal plain is the modern-sediment example of the process. Geologists have studied the coastal plain for years to unravel the secrets of these sediments and their processes in order to use this information to interpret and locate ancient deposits of oil, gas, water, uranium, coal, and other resources.

Cross section, I-10 between San Antonio and Houston.

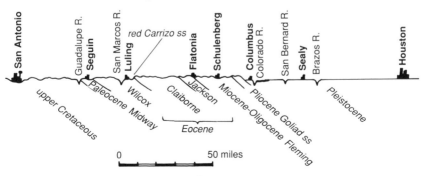

Point bar at river bend. San Bernard River at Interstate 10 looking north.

I-10 crosses major Texas rivers responsible for much of the coastal plain deposition, namely the Brazos, Colorado, and the smaller San Bernard.

At bridge crossings you will get a quick look into the gorges of these rivers. Note especially that they are incised into the coastal plain about 50 feet. This incision occurred during the Pleistocene Ice Age when sea level was much lower, climate was wetter and rivers carried more water and sediment load, on the average, than they do today. The rivers balanced all these factors by downcutting.

Meander bends, sand and gravel bars, point bars, steep outer banks, low inner banks — all features of meandering streams—can be seen from I-10 where it crosses the Colorado, Brazos, and San Bernard rivers. The San Bernard particularly has nice gravel bars seen when water level is low; construction gravel is mined in pits on the south side of the freeway at the Colorado River.

Between Columbus (Colorado River) and San Antonio the topography is hilly and rolling in sharp contrast to the flatness of the coastal plain eastward toward Houston. This marked change is noted just west of the Colorado River crossing. With one exception no outcrops of rock are visible along this stretch of freeway, but the topography is a key to the underlying geology. I-10 runs perpendicular to a series of northeast-southwest oriented sand ridges and intervening clay swales produced by differential erosion of inclined Tertiary-age rocks. These rocks are the product of sedimentation of wedges of sand and clay built into the Gulf of Mexico over the last 50 million years.

Look carefully for the one outcrop of red (Eocene-age) sandstone on a ridge on the south side of I-10, two miles east of the Luling exit at the Gonzales County line sign. The red soil color is common in high rainfall areas where the chemical weathering of ground water leaches out all the calcium and potassium, leaving behind oxidized iron (rust!). Moreover, the Carrizo sandstone here also contained abundant iron to begin with! Iron-rich sandstone blocks can be seen in Palmetto State Park, just south of I-10.

Between Seguin and San Antonio, I-10 rolls along on terrain underlaid by upper Cretaceous mudstones and siltstones (Taylor, Navarro, and Midway groups of rocks). These rocks are quite soft and easily eroded, so outcrops along this stretch are virtually non-existent. The town of Seguin rests on a high alluvial terrace of the ancestral Guadalupe River—one of many terraces that parallel the courses of modern-day Texas rivers.

Interstate 10
Houston—Beaumont—Orange
112 miles

This trip covers the complete story of petroleum in Texas. At one end is Houston, where the modern-day petroleum industry is headquartered, while at the other end is Beaumont where it all started, and in between, the road traverses depositional environments which tell the tale of how petroleum is formed.

Houston is, of course, the nerve center for the worldwide petroleum industry. Virtually every major, and many smaller, world petroleum corporations are either headquartered or have a presence here. It could also be said that Houston is the global capital of geology, based on the fact that more geologists reside and ply their trade in Houston than any other city in the world. The worldwide search for petroleum, as well as the financing, production, refining, shipping, pipelining, supplying, and marketing of the world's principal energy material in the 20th century largely emanates from Houston.

While Houston is the modern-day Goliath of petroleum, the town of Beaumont, 90 miles away, is the earth-mother, where the petroleum age dawned in 1901. On a cold day of that year, January 10th to be exact, at 10:30 in the morning, the first great American gusher "roared in like a shot from a heavy cannon and spouted oil a hundred feet over the top of the wooden derrick out on the hummock that the world would soon know as Spindletop. This oil discovery changed the world. Before Spindletop, oil was used for lamps and lubrication. The famous gusher of Captain Anthony F. Lucas changed that. It started the liquid fuel age, which brought forth the automobile, airplane, the network of American highways, improved railroads and marine transportation, the era of mass production, and untold comforts and conveniences."*

Spindletop itself is "a little knob of land rising out of a swampy prairie in the southeast corner of Texas," south of the town of Beaumont, at the newly named (in 1901) town of Gladys City, Texas. The hill is the surface expression of a large subterranean salt dome, which trapped enough oil to produce 100,000 barrels per day in the initial flow from the Spindletop discovery well. Beaumont became an immediate boom town, and large and famous oil companies (Gulf, Texaco) got their start here. All the elements of a burgeoning petroleum industry existed in the Beaumont—Orange area in 1901. Clipper ships called in at the Port of Beaumont in the 1800's, so a shipping network was well established. Timber and shipbuilding were major industries at Orange, which had ready access to the Gulf of Mexico via the Sabine River. Wooden derricks and other oil field equipment could thereby either be locally built or easily shipped in from elsewhere.

The Spindletop Museum at Lamar University in Beaumont features artifacts and documents from the early oil days and a film explains the Spindletop discovery. The nearby Gladys City—Spindletop Boom Town is a reconstructed turn-of-the-century oil town complete with wooden derricks, early oil field equipment, woodenfront stores, a blacksmith shop, saloon and post office.

Between Houston and Beaumont, Interstate 10 rolls along on the flat terrain so typical of the Gulf Coastal plain. Eighteen miles from Houston, the highway crosses the floodplain of the San Jacinto River, and twenty miles farther it crosses the wide, swampy and lakestudded valley of the Trinity River. Both rivers empty into Trinity Bay, which is at the upper end of Galveston Bay, the entire bay complex occupying drowned mouths of both the Trinity and San Jacinto rivers. Once these rivers extended many miles to the south, even beyond the present-day sea coast at Galveston, when sea level was lower during the Ice Age. In the last few thousand years, sea level has risen, and literally drowned the river mouths to form bays. A rich mix of freshwater and saltwater organisms occupy the waters of the shallow bays, lakes, swamps, and channels. An equally organic-rich

*quotes from Clark, J.A., and Halbouty, M.T., 1952, Spindletop, Random House, N.Y., 306 pp.

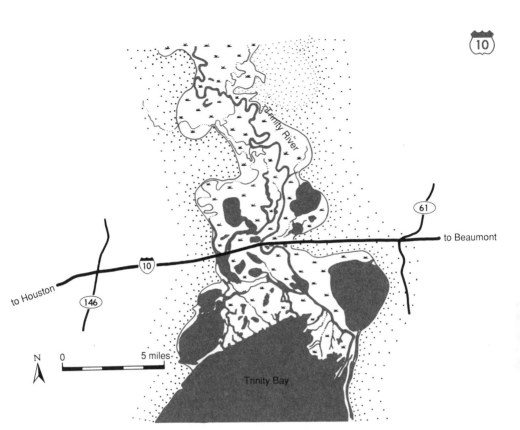

Drowned mouth and delta of Trinity River on I-10 east of Houston.

near- tropical plant assemblage grows in the wet climate (58 inches of rainfall per year) of these environments and contributes to the endemic soup of organic debris.

Keep this scene in mind as you get a perhaps too-quick look at the Trinity valley. The valley is a modern day model for environmental conditions that have existed along the Gulf for many millions of years. And, these very environments, buried deeply in the continually sinking sedimentary wedge of the Gulf are responsible for the extensive petroleum deposits in this area. The organic materials are the sources for the petroleum. Buried to many thousands of feet, temperatures become hot enough to cook this organic material to form oil and gas, while the sands in the buried stream channels and beaches become the reservoirs as the newly cooked oil seeks to occupy the tiny, myriad spaces between the sand grains. Salt domes pushing up the thick layers of organic mud form the structure for the indigenous hydrocarbons to migrate into and be entrapped. Take another look at the Trinity valley—the swamps, lakes, channels, the surrounding plants and mud—and picture it all buried deep beneath the surface. If you now envision oil and gas forming, your mental image is perfect!

This way of thinking illustrates a fundamental geologic principal: the present is the key to the past. This principal, also called "uniformitarianism," links modern environments, as seen along Interstate 10, to ancient rocks, in a way that allows the interpretation of the environment in which ancient sediments were deposited. By looking in detail at modern environments, the geologist can examine an outcrop or a sample taken from deep beneath the surface and tell what the conditions were like in ancient geologic times. The present is the key to the past!

Interstate 35
Temple—Austin—San Marcos
99 miles

Between Temple, the wildflower capitol of Texas, and Austin, the state Capitol, Interstate 35 runs northeast-southwest, paralleling the edge of the Hill Country, which lies to the west of the highway. The road is mostly on the flat terrain that lies at the foot of the Balcones ("The Balcony") escarpment, but south of Temple, on the skyline to the west the elevated hills marking the edge of this escarpment are readily apparent. Between Temple and Georgetown, I-35 follows a band of lower Cretaceous limestone. The freeway swings eastward between Georgetown and Austin, encountering younger, upper Cretaceous rocks. Here and there a glint of white limestone can be seen in stream cuts beneath bridges, or in small roadcuts, but you have to look carefully to find these few, rather poor exposures.

At Georgetown, Inner Space Caverns is an underground cave open to visitors. The cave formed on the edge of the uplifted limestone block

Austin chalk roadcut south of Austin on Interstate 35.

Limestone quarry and fence. Texas Crushed Stone Company, south of Georgetown on Interstate 35.

along the Balcones fault system, as subterranean groundwater etched its way through the cracks in the limestone. Another cave, formed in a similar setting, is Wonder Cave, at San Marcos (see Texas Caves in the Central Section), while Aquarena Springs, a popular and entertaining site in San Marcos, is fed by spring water emanating from the upland limestone terrain.

South of Georgetown (Exit 257 - Westinghouse Road) is a huge limestone quarry west of I-35. The Texas Crushed Stone Company mines Cretaceous Edwards limestone here, mainly for road aggregate. Blocks of limestone are lined up along the feeder road to form a stone fence, where you can get a close look at the rocks. The Edwards limestone is the main rock unit that forms the Edwards Plateau west of Austin and north of San Antonio. Springs, caves, and sinkholes are common in the Edwards limestone, and the water in the subterranean labyrinth of the Edwards supplies San Antonio's drinking water.

The Interstate highway crosses the Colorado River at the south edge of downtown Austin (see separate Austin segment), though outcrops are not seen in the river banks near I-35. However, south of Austin, watch for a nice exposure of white Austin chalk, on the east side of I-35 at Slaughter Creek, about six miles south of the Texas 71 crossover. Between Austin and San Marcos I-35 rides on flat Cretaceous terrain. Past Onion Creek the road surface is on upper Cretaceous Taylor and Navarro mudstones, which are soft and do not stand up as natural outcrops, so this group of rocks is not easy to observe from the highway. North of San Marcos the Balcones escarpment is quite obvious across the flat fields west of the freeway.

Geologic map of Austin.

AUSTIN

Austin is the gateway to the Hill and Lake Country west and northwest of town. The Texas state capital city straddles the line between two geologic provinces—to the west is the uplifted plateau of Cretaceous limestones and exposed Precambrian Llano uplift, while the lowland Black Prairie country stretches eastward for miles. Sharply dividing these two distinct topographic terrains is the Balcones fault which forms a northeast-southwest line through the western half of Austin, creating the Balcones escarpment. The escarpment has been the location for colonization and towns throughout Texas history because of the natural resources associated with the fault—plenty of natural building stone outcrops on the escarpment, water pours from springs, timber can be gotten on the wooded hills, and the adjacent fertile prairies support a strong agricultural economy.

The Spanish explorer, Bernardo de Miranda, in 1756 named the escarpment "Los Balcones", meaning "balconies", which describes quite well the stair-step, balcony-like topography rising above the plains.

At Austin, the offset along the Balcones fault—really a fault zone, not a single fracture line—is about 1,200 feet, while 600 feet of displacement takes place along the Mt. Bonnell fault, the westernmost

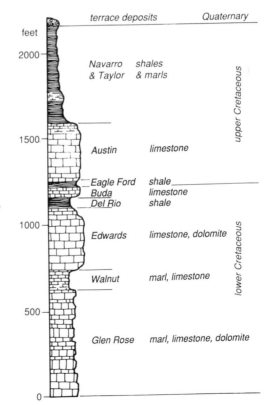

Column of rocks in the Austin area.

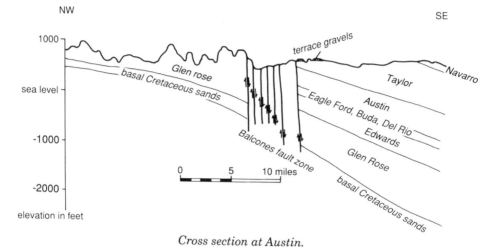

Cross section at Austin.

break of the fault zone. The offset can be seen on the geologic map and cross section. Note how the lower Cretaceous Glen Rose formation is at the surface west of the Mt. Bonnell fault, while east of the fault zone younger rocks are at the surface, overlying the Glen Rose which here is buried 1,000 feet below.

The rocks around Austin are mostly limestone, but dolomite, clay, basalt, tuff, sandstones, and river gravel are also present. Limestone ledges are commonly seen around the edges of Lake Travis. They are part of the Glen Rose formation, famous for its dinosaur tracks in many places in Texas. The Edwards limestone is another prominent, thick unit seen in roadcuts and outcrops northwest of town along U.S. 183. A band of white limestone, called the Austin chalk, obviously named for outcrops here, cuts a swath down the center of the city. Exposures of Austin chalk are seen in many places in town. You can get a quick highway look at the Austin chalk in a roadcut on Interstate 35 at Slaughter Creek about five miles south of the city. East of Austin, in the prairie country, softer shales, sandstones, and marl of the upper Cretaceous Taylor and Navarro formations aren't hard enough to resist erosion, so outcrops of these rocks are not so common.

Springs emanate from limestone aquifers all up and down the length of the Balcones escarpment. One good example in Austin is found at Barton Springs, a favorite swimming and picnicking site south of the Colorado River. The water comes to the surface at Barton

Springs where a fault has offset a porous unit within the Edwards limestone, which is honeycombed with caves and solution cavities.

Pilot Knob, southeast of city center on U.S. 183 near Bergstrom Air Force base, is a cluster of small rounded hills which expose the core of a Cretaceous volcanic explosion crater. Black, fine-grained nephelinite—a volcanic rock related to basalt, along with reddish tuff or volcanic ash, and pyroclastic debris blown from the volcano, make up this fascinating assemblage of rocks on Pilot Knob. About 80 million years ago this area of Texas was on a shallow marine shelf where limestone rock was deposited by the assemblage of lime-secreting organisms that lived in the sea at that time. As the Gulf of Mexico continued to form, fractures broke the earth's crust in a line following the Balcones fault zone, and hot, molten lava rose through the fractures from deep in the earth's mantle. As the lava pushed upward through the sediments and sea water, explosions driven by steam rocked the sea floor. Craters formed around these explosion vents, one of which is Pilot Knob. Ash, debris, and lava filled the crater; eventually a dome extended above the sea floor. When volcanic activity ceased and the lava had cooled, reefs developed around the mound, and even beach rock formed as waves lapped the margins of the mound. Such reef-beach rock can be seen at the falls in McKinney Falls State Park just north of Pilot Knob.

Finally, the mound was covered as the sea bottom subsided throughout the Cretaceous period, and clay and the Taylor marl were deposited over the top of the mound. About ten million years ago, in Miocene time, uplift occurred along the Balcones fault, and the Cretaceous rocks were elevated, exposed, and eroded. The harder volcanic rock resisted erosion a little better than the limestone, so

Barton Springs in Austin.

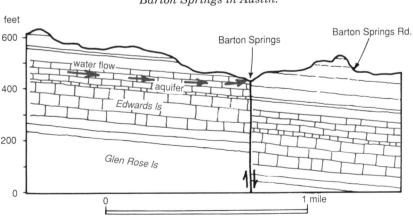

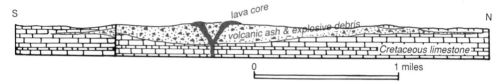

Pilot Knob explosion crater.

Pilot Knob at 710 feet above sea level, today stands 180 feet above the surrounding terrain. Pilot Knob is perhaps the best known of dozens of volcanic craters and mounds that string out along the Balcones fault zone. The largest concentration of Cretaceous volcanic mounds is found around Uvalde to the southwest of Austin.

A pleasant view of the city and surrounding countryside can be gotten from Mt. Bonnell which rises 775 feet above sea level. Glen Rose limestone, elevated west of the Mt. Bonnell fault is the bedrock around Mt. Bonnell.

The state capitol building is, appropriately, a geologic monument in itself. Edwards limestone was quarried locally in Travis county for part of the stonework, whereas the Texas pink granite came from Granite Mountain near Marble Falls, where beautiful granite building stone is still actively mined. Construction on the capitol began in 1882 under the direction of architect E. E. Meyers of Detroit, Michigan. The edifice was completed and dedicated in 1888.

Geology displays are featured in the Texas Memorial Museum on the University of Texas campus, while a treasure-trove of Texas geological literature can be purchased from the Bureau of Economic Geology, Balcones Research Center, Building 130, located at 10100 Burnett in northwest Austin.

For readers interested in a detailed guidebook to the geology of Austin, the following is recommended:

Guide to Points of Geologic Interest in Austin, by A. R. Trippit and L. E. Garner, 1976, Guidebook Number 16, Bureau of Economic Geology, the University of Texas, Austin. 38 pages, with colored map.

Interstate 45
Houston—Galveston
45 miles

Contrary to popular belief, as much geology is to be seen in the Houston—Galveston corridor as anywhere in the state. However, if you have just returned from a trip to Big Bend National Park or Palo Duro Canyon or the rocky Hill Country, this statement is undoubtedly stretching your credibility quotient about now, as you peer across terrain so flat that the highest topography is found on freeway overpasses! But the comparison between the geology of Houston—Galveston and the geology of Big Bend Park is a comparison between two major aspects of the study of geology. The rocks in Big Bend tell a story of events, now frozen in time, which happened in distant past millenia, while the events themselves are taking place right now around Houston—Galveston. Here, to be observed in "real time", are active faults, subsiding land, flowing-depositing-eroding-flooding rivers, hurricane-blasted barrier islands, tidally-washed bays, beaches

Subsurface faults, oil fields, and salt domes (dotted circles in black) in the Houston area.

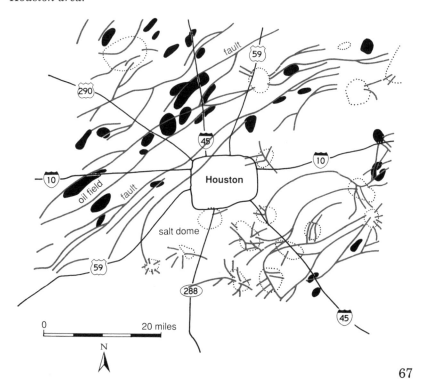

where sand drifts 'longshore,' driven by pounding surf and waves, and swamps and marshes where organic-rich muds belch bubbles of fermented gas. This gas foretells of hydrocarbons to come in some distant future millennium when all the sediments in today's Houston –Galveston environments will be buried, in their turn, under thousands of feet of sediment. Modern environments, and the dynamic interplay of surface geologic forces on those environments, is a major subject of geologic investigation; the Houston–Galveston area is a place where "action geology" can be observed.

The Houston metropolis rests on the flat surface of the coastal plain, which is at the top of a giant wedge of mud and sand deposited into the Gulf of Mexico during the Ice Age (Pleistocene age) by ancestral Texas rivers. These floodplain, river and delta muds support a dense growth of semi-tropical vegetation that thrives in the mild climate and 48 inch annual rainfall.

Subsidence in the Houston–Galveston area.

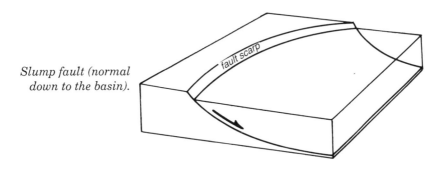

Slump fault (normal down to the basin).

Houstonians do not live on a stable surface, however, for numerous active faults criss-cross the city, which surprises many people who associate faulting with mountains in places like California. But the truth is that the giant mud wedge is naturally sliding into the Gulf of Mexico along long, curved faults, while at the same time sinking as the mud compacts under its own weight. As the mud compacts, water is squeezed out, so Houston rests on the nation's third largest aquifer, from which the city merrily pumped groundwater from 200 wells until the 1960's, when the realization came that the entire metropolitan area was sinking at an accelerated rate! For example, the area around the San Jacinto Monument sank six feet between 1900 and 1964. Removal of oil and gas early in the century also contributed to local sinking problems. The water pumping was stopped, drinking water now comes from surface ponded water in Lake Livingston and Lake Houston, and less and less oil and gas is removed from depleting reservoirs, so the rate of subsidence in Houston has decreased significantly. But the natural subsidence from compaction will continue and natural sliding along faults can't be stopped, so hurricane and thunderstorm flooding can be expected to become more severe in subsiding areas.

Don't leave Houston without visiting the Museum of Natural Science in Houston's Hermann Park. Excellent displays feature dinosaurs, geology, petroleum geology, and a fantastic mineral collection.

Between Houston and Galveston, the Gulf Freeway (Interstate 45) rides on the flat surface of Pleistocene rivers and delta muds. Note the intermittent swamps, incised bayous ("creeks") and rich flora along the way. When this landscape is buried under a few thousand feet of sediment, which it will be some day, it will be a source for new oil and gas—in a few million years.

Movement on the Longpoint fault in northwest Houston has caused a three foot drop in the street and cracked this driveway. The dotted line shows approximate fault trace with up and down sides indicated.

Cracks in the brick wall where a house is located right on the Longpoint fault.

70

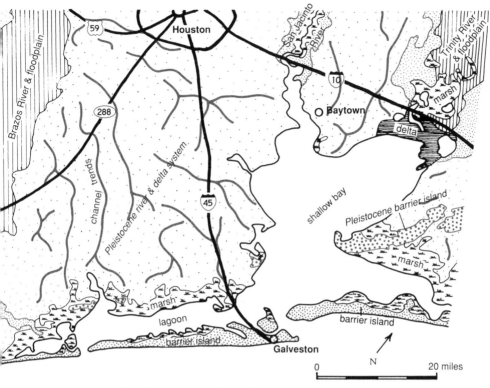

Environments in the Galveston–Houston area.

The freeway follows the west shore of Galveston Bay, though you can't see the bay from the road. The bay, of course, occupies the drowned mouths of the San Jacinto and Trinity rivers. The drowning occurred when sea level rose nearly 400 feet as climate warmed and the world's vast glaciers melted after the Ice Age. Bay waters are only about 12 feet deep, so to accommodate shipping a ship channel was dug early this century making Houston the nation's third largest port today.

Bay waters circulate mainly in a counterclockwise pattern, driven by dominant winds blowing out of the southeast, and by low tidal surges which enter at the passes on either end of Galveston Island. Oyster shell banks are common in Galveston Bay, which is only one indicator of the rich biological productivity that characterizes these shallow bays.

Watch for NASA Road-1, about 13 miles south of Houston, where the Johnson Space Center is located. In addition to Mission Control, the Museum, and outdoor rocket displays, be sure to visit the Lunar Sample Laboratory, Building 31A, where the largest collection of

71

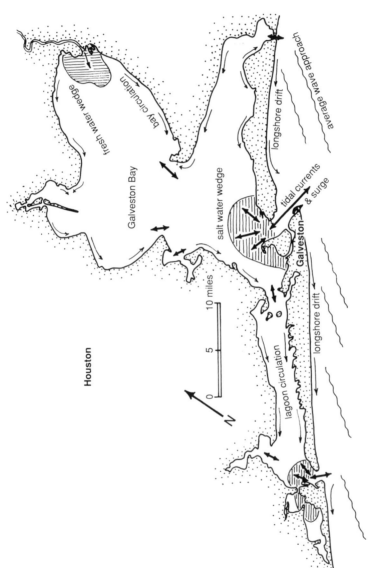

Tides, waves, and water circulation in the Galveston Bay area.

72

moon rocks is stored and displayed and where analytical equipment for studying these important "geologic" samples is located.

Tall refinery columns rise like a metallic forest east of the freeway at Texas City, where a large percentage of the nation's gasoline and petrochemicals are refined. Texas crude once supplied the majority of the feedstock for the Texas City refineries, but now more and more crude oil comes from foreign sources, as the United States' reserves dwindle.

The shallow lagoonal waters of West Bay come into view as the freeway nears the causeway bridge to Galveston. Patches of marsh grass form circular patterns seen to the west of the road. The grass flats and salt marshes border both the mainland and barrier island sides of the lagoon, forming important habitats for birds and underwater dwellers.

The city of Galveston on Galveston Island is located on a Karankawa Indian site, and is the place where the Spanish explorer, Cabeza de Vaca, was storm-tossed from a shipwreck in 1528. The pirate Jean Lafitte started the first European settlement on the island in 1817, and since then, Galveston has attracted sun worshipers with its 32 miles of sandy beaches.

In 1900 the entire island was inundated by a powerful hurricane, and tragically, 6000 lives were lost. Shortly thereafter (1902) the Galveston seawall was built, houses were jacked-up and the city was elevated twelve feet with landfill. Since then the ten-mile seawall has

Patterned grass flats in West Bay along Interstate 45 to the west, just before crossing the causeway from the north.

Swash along Galveston Beach. Seawall in the background.

West end of Galveston Island. Beach and low, vegetated dunes.

Bird and crab tracks on wind-rippled dunes.

Oyster shell hash on the lagoon side of Galveston Island.

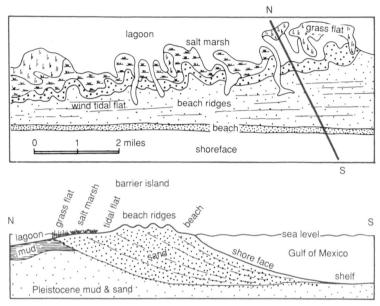

Map and cross section of Galveston Island environments.

proved effective, as Galveston has successfully withstood numerous hurricanes since 1900.

A series of long stone jetties are part of the Galveston beach protection system. The large blocks in the jetties are native Texas granite, mined from the Precambrian Town Mountain granite in quarries northwest of Austin at the town of Marble Falls (see Central Texas section). This same granite is the principal building stone in the Texas state capitol building in Austin.

To see the natural environments of Galveston Island, drive west on Seawall Boulevard, beyond the end of the seawall itself. Note the rather flat sandy profile to the island. Because of high vegetation growth, much sand is tied up by plants, so not a lot of free sand is available to build dunes, though low dunes are found behind the beach in many places. Hurricanes also play a vital role in flattening the profile of the island. A drive north from Seawall Boulevard on any number of back streets will take you across the beach ridges, and give a view of the muddy, shelly banks, salt marshes, and grass flats that characterize the lagoon-side of the island.

Galveston Island State Park, west of the city, preserves a nice segment of natural beach—beach ridge—salt marsh and grass flat environments.

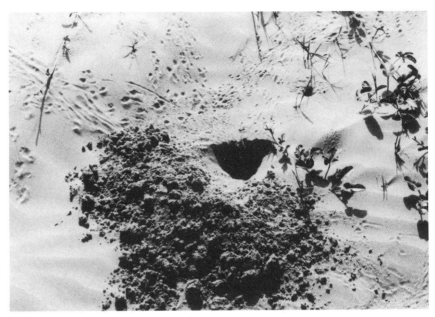

*Ghost crab burrow and tracks are usually found on the back
beach against the dunes.*

Common shells and animals found on Galveston Beach:
1. Cockle clam fragment
2. Texas venus clam
3. Sand dollar fragment
4. Common oyster shell
5. Snail fragment
6. Tube of a tube-building worm
7. Whelk snail fragment
8. Speckled crab
9. Speckled crab
10. Purse crab
11. Cockle clam
12. Pen shell clam
13. Snail borings in shells
14. Texas venus clam

At the east end of Galveston Island at Apffel Beach, wind-blown dunes display wind ripples and cross-bedded internal structures, while a wide expanse of vegetated back-dune flats host swampy pools where waterfowl abound.

Poking around any of the Galveston beaches is great fun, and shell searches and hunts for ghost crab holes can while-away many pleasant hours.

Interstate 45
Houston—Centerville
111 miles

Between Houston and Centerville, Interstate 45 crosses the upper coastal plain and transition onto the bordering upland piedmont. Northward from Houston, the freeway rolls along on the flat floodplain of Ice Age (Pleistocene) sand and clay deposits, crossing the west fork of the San Jacinto River a few miles south of Conroe. The northern edge of the coastal plain extends to beyond Willis, nearly to New Waverly, about 50 miles north of Houston. The first, low-rolling hills encountered there mark the presence of somewhat harder, underlying Miocene sandstones and shales.

Between Huntsville and Madisonville, an older, Eocene set of sandstones crosses the highway at nearly right angles, creating an even more noticeable up-and-down topography. Roadcuts are rare, but you can see a sandy exposure here and there along the way.

The hilliness south of Centerville is mainly due to a ridge of Eocene Sparta sand, which is a quartz sand that weathers yellow and brown.

All the Pleistocene to Eocene sand and clay deposits encountered along this section of highway, are part of the thick Tertiary wedge of sediment which Texas rivers have been dumping into the Gulf of Mexico for nearly 60 million years. The wedges of sediment tip southward; units exposed in this area lie deep under the surface below Galveston.

Many of these sedimentary rocks were exposed in freeway roadcuts when the highway was being built, but are now hidden behind the grassy slopes bordering the road.

Interstate 45 northward from Centerville to Dallas is continued in the Northeast section.

U.S. 59
Houston—Victoria
120 miles

The drive between Houston and Victoria is entirely across the flat expanse of the coastal plain. Rich soils from overbank flood deposits and a rich vegetation produce much organic humus in the humid climate. Two of Texas' major rivers are crossed in this stretch—the big bend of the Brazos River at Richmond/Rosenberg and the Colorado River at Wharton. Views from the bridges are good of sand bars on the inside bends and the steep opposing outside banks of these rivers. The Colorado and Brazos rivers slice diagonally across the entire state, heading in the high plateau of the northwestern Texas Panhandle, both emptying into the Gulf of Mexico a mere twenty miles apart. The color of these rivers varies from clear to brown and even red (particularly the Brazos), depending on sediment load, which in turn is associated with upstream rainfall and erosion. The Brazos generally runs redder because it crosses and erodes red Permian rocks in northcentral Texas, whereas the Colorado tends to run clearer in its lower reaches largely because many dams trap sediment upstream.

The highway crosses many other small streams and creeks which roughly parallel each other as they flow south-southeast toward the Gulf.

U.S. 59
Houston—Lufkin
117 miles

This north-south trek is another transect across the upper coastal plain and onto the low hills of the adjacent piedmont which is underlaid by sandy bands of Ice Age (Pleistocene) and older Tertiary sedimentary rocks. The 40 miles north of Houston to Cleveland is on the flat coastal plain, which is broken only by crossings of the west fork and east fork of the San Jacinto River. Tan sandy soils that predominate here are covered by extensive pine stands, which love such soils. Orange-tan sandy soils and poor roadcuts are seen around Shephard, about 50 miles north of Houston, where the Trinity River creates a bit of topography in the Ice Age deposits of the Willis formation. The more rolling topography north of the Trinity River crossing to Livingston marks the edge of the Tertiary piedmont, where Miocene orange-tan sands and soils are found.

The town of Moscow has a quarry in tan Miocene sand. It also has Dinosaur Gardens, a local dinosaur exhibit which offers ten full scale dinosaur models and a nature trail. Five miles north of Moscow is a roadcut in flat-lying, light gray sandstone which represents river channel deposits of Miocene age.

The highway crosses the Neches River, about 100 miles north of Houston, between Corrigan and Diboll. The abundant pines growing on the sandy soils atop Miocene sandstones in this area support an active timbering industry. Diboll is the center for forest products manufacture.

Rolling hills around Lufkin are on older, Eocene sandstones. Note how the topography increases as the underlying formations get older. The sandstones harden with age, becoming a little more resistant to erosion, so hills stand higher. East of Lufkin is the large Sam Rayburn Reservoir, which is a dammed segment of the Angelina River, a tributary of the Neches River, which ultimately feeds into the Sabine River at Port Arthur.

[U.S. 59 northward from Lufkin to Kilgore is continued in the Northeast section.]

U.S. 290
Houston—Brenham—Austin
152 miles

This highway affords a nice drive between Houston and Austin. The road traverses pleasant rolling hill country as it crosses sandy ridges and shale-based valleys cut into the tilted package of Tertiary sediments which lie adjacent to the coastal plain.

U.S. 290 heads northwest out of Houston in a straight line for nearly 20 miles from its junction with West Loop 610, travelling on the flat surface of coastal plain sands and muds. At Cypress, U.S. 290 makes a bend toward the west; three miles farther is an abrupt change in topography where the coastal plain ends and Tertiary uplands

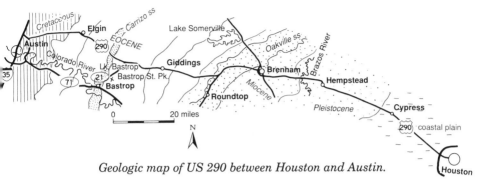

Geologic map of US 290 between Houston and Austin.

begin. The road climbs a noticeable hill, leaving the flat coastal plain behind; from here to Austin the topography is markedly rolling.

Hills around Prairie View and Hempstead are held up by Pleistocene sandstones of the Willis formation, into which the Brazos River has entrenched itself. The road crosses the Brazos Valley a few miles west of Hempstead, then climbs through river-edge topography to the high hill on which Brenham is built. Sandy ridges west of Brenham form the Oakville escarpment, which extends for many miles to the southwest as a recognizable topographic ridge. The Miocene Oakville sandstone is hard enough to preferentially resist erosion, and stand up as a ridge. A few miles northwest of Brenham, on Texas 36, is Lake Summerville, a popular recreation site. Sandstone outcrops around the lake are mainly Eocene in age.

Southwest of Brenham, on Texas 237, is the historic town of Roundtop, where the University of Texas Winedale Seminar Facility is located. Outcrops of Oakville sandstone are fairly well exposed on the ridgetop in and around Roundtop. Historic buildings, stone fences, and foundations in both Brenham and Roundtop are built of the native sandstone.

Between Brenham and Elgin, the road bobs up and down as it traverses progressively older Tertiary sandstone and shale. Younger Miocene bands are found around Brenham, whereas older Eocene rocks surround the town of Elgin. Roadcut exposures are not common on this vegetated stretch of roadway, but if you watch carefully, you will see a sandstone ledge poking out here and there from the roadside grass.

Hills of strikingly bright red sandstone a few miles west of the junction of U.S. 290 and Texas 21 and at Bastrop State Park southwest of U.S. 290 on Texas 21, are exposures of Eocene Carrizo sandstone. This red band of iron-rich sandstone is distinctive for many miles across southeast and southwest Texas. Pine trees are localized on the sand ridges along Texas 21. Pines prefer the water-holding sandy substrate and are good indicators therefore of sandy soils.

Between Elgin and Austin, U.S. 290 rides on upper Cretaceous-aged rocks. These marly limestone units are not exposed very well along this segment, but look for the glint of white limestone fragments, plowed up to the surface in farm fields.

Texas 36
Richmond/Rosenberg—
Damon Mound—Freeport
57 miles

From U.S. 59, Texas 36 heads south across flat coastal plain terrain, where cotton, sorghum, and formerly much sugarcane, are grown on the rich brown soils. After passing through the small communities of Needville and Guy, a large hill looms on the southwest skyline near the town of Damon. Looking oddly out of place standing 83 feet above the flat coastal plain—which is quite a landform here—Damon Mound is one of the best surface expressions of a salt dome to be seen on the Gulf Coast. A quarry carves out the northwest corner of the mound, where the caprock of Oligocene-age coral reefs (Anahuac formation) is mined for building stone, crushed gravel, and at one time, sulfur. A variety of corals, pecten clams, other clams, and one-celled foraminifera are the reef-building organisms found here. Oil wells once surrounded the mound, outlining the underlying salt dome; the wells were drilled in the teens and twenties. Over ten million barrels of oil were produced from the Damon Mound oil field between 1917 and 1952.

The salt dome has pushed its way upward to create the mound, from the Jurassic Louann salt layer thousands of feet beneath the surface. Interestingly, Pleistocene-age flank rocks are tilted, which is evidence for upward migration of salt quite recently in geologic time.

Damon Mound was a favorite campsite for Karankawa Indians, as evidenced by their burial sites, arrowheads, pottery, and stone implements found on the mound. The town of Damon, on the east side of the mound, began in 1831 with the construction of a blacksmith shop out of limestone from a natural outcrop. The town cemetery has a number of 1830's headstones, and is a registered Texas Historical Commission Landmark.

The road continues south from Damon through West Columbia, Brazoria, and Jones Creek, all on coastal plain mud and sand, before it crosses the Brazos River just north of Freeport. At Freeport, the intercoastal waterway separates the coastal beach from the mainland. Bryan Beach, south of Freeport, is a popular undeveloped state park which sun lovers enjoy, while Brazoria National Wildlife refuge, north of Freeport, hosts thousands of birds which enjoy the preserved wetlands. Highway FM 3005 follows the barrier island chain, and connects Freeport and Galveston.

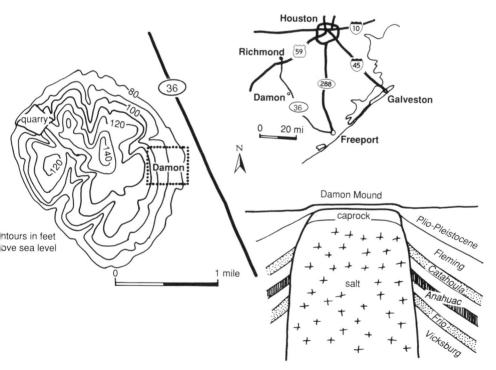

Damon Mound is a striking surface expression of a salt dome on the otherwise flat coastal plain southwest of Houston.

Southwest Texas — Lower Gulf Coast

Southwest Texas
Lower Gulf Coast

Interstate 35
San Marcos—San Antonio
38 miles

The interstate between San Marcos and San Antonio is built on upper Cretaceous claystones and marls, which don't form very good outcrops because of their easily erodable nature. Just to the west is a line of skyline ridges which are upheld by hard, resistant Edwards limestone, about 25 million years older than the claystones and marls on which I-35 is constructed. The older limestone is found higher than the younger claystones marls because one of the main faults of the Balcones fault system lies between the highway and the skyline ridge. It was along this fault that the limestone hills were uplifted in Miocene time, about ten to twenty million years ago. This uplift raised the Hill Country and Edwards Plateau northwest of San Antonio to about 2,000 feet above sea level. This movement on the Balcones fault occurred about the same time that the Colorado Plateau in Four Corners area in the southwest United States was elevated and the Rio Grande rift in New Mexico opened up. Apparently a large part of western North America underwent simultaneous uplift and faulting during Miocene time.

Springs emanate from the limestone along the fault line, and Comal and San Marcos springs are the largest in the southwestern United States. These springs are partial outlets for the water in the immense Edwards aquifer, which supplies San Antonio's drinking water. Comal and San Marcos springs are the sources for the Comal and San Marcos rivers in this area.

Huge limestone quarries are seen west of the road between San Marcos and San Antonio, one at the Comal County line a few miles south of San Marcos, another south of New Braunfels at Rueckle Road Exit 184, and yet another at Exit 177 on the Guadalupe County line nearer San Antonio. A lot of cement, road gravel and aggregate from these quarries has fueled the economy of this area. The cement and aggregate is mainly shipped to coastal cities such as Corpus Christi and Houston where limestone and rock is in short supply. The

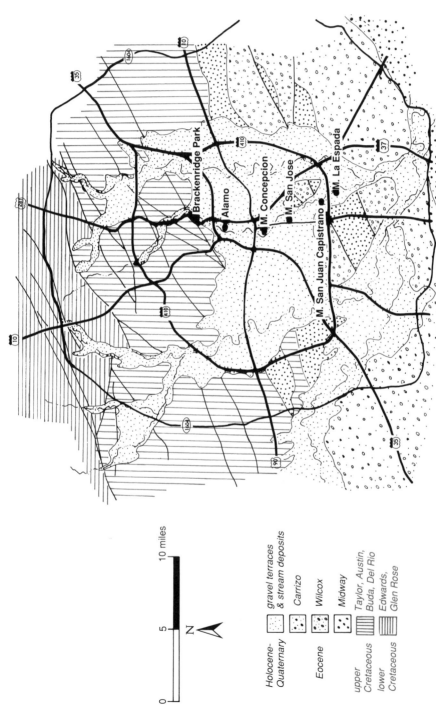

Geologic map of San Antonio.

10 miles

5

N

0

Holocene-
Quaternary

gravel terraces
& stream deposits

Eocene

Carrizo

Wilcox

Midway

upper
Cretaceous

Taylor, Austin,
Buda, Del Rio

lower
Cretaceous

Edwards,
Glen Rose

Brackenridge Park

Alamo

M. Concepcion

M. San Jose

M. La Espada

M. San Juan Capistrano

Edwards limestone is the main stone quarried in these and similar quarries at many places along the Balcones escarpment north of San Antonio.

SAN ANTONIO

San Antonio is a city built on Texas geology. It not only is a center for geologists who make their living in the state looking for geologic resources—oil, gas, water, and minerals, but it sits astride a fundamental juncture in the state's geology. The Balcones fault system, nearly 30 miles wide, slices northeast–southwest right through the city, separating the high, upthrown Cretaceous limestone terrain of the Edwards Plateau and Hill Country north of town from the flat, lowland sandstone and mudstone terrain of the Coastal Plain south of town.

The Balcones fault "popped up" about ten million years ago (Miocene), elevating the Edwards Plateau nearly 2,000 feet above sea level. The fault follows the edge of the old buried Ouachita Range, which means both the fault and the range lie along a deep-seated suture in the Earth's crust. The fault is not a single break in the rocks around San Antonio, nor a single line on the geologic map, but rather a zone of stair-stepping faults. As you drive north through San Antonio, and particularly north on Interstate 10—the up and down topography, and knolls of limestone north of town are the result of this pattern of fault slices and intervening fault blocks. Along U.S. 281, this block pattern also shows up quite well, especially around Encino Park.

Cross section showing recharge of porous rocks at the surface and flow of groundwater beneath the surface at San Antonio.

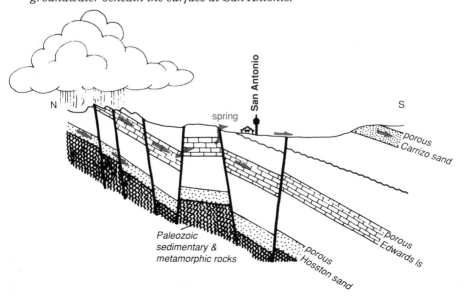

Wall of Cretaceous Austin chalk limestone in old Alamo cement quarry, now the Japanese Tea Garden in Brackenridge Park.

Close up view of fossil hash in Austin chalk limestone in Brackenridge Park. Long, white pieces are fragments of large, Cretaceous clam, Inoceramus.

Blocks of limestone in walls of the Alamo possibly came from quarries near Brackenridge Park.

Cretaceous limestone, particularly the upper Cretaceous Austin limestone, has been quarried in San Antonio for centuries—mainly for building stone. The Austin is not as hard as the Edwards limestone, nor as riddled with solution holes, so it is easier to mine and makes a more uniform building stone. One of the best places to see limestone close-up in a quarry is at Brackenridge Park, where the Japanese Tea Garden is built against the backdrop of limestone quarry walls. The Spanish quarried Austin limestone from this location in the 1700's and limestone blocks in the walls of the Alamo may have come from here. Many other historic buildings in San Antonio are also faced with Austin limestone blocks. The outer walkway of the Japanese garden winds along the quarry wall, where typical Cretaceous oyster shells and giant clams (Inoceramus) are abundantly evident. Remnants of the nineteenth century Alamo Cement Company mill are also part of Brackenridge Park.

Another historic geologic site is San Pedro Park. Just off I-35 at the San Pedro exit, and north on San Pedro Avenue, San Pedro Park hosts a natural spring which bubbles fresh water from Austin chalk. The springs rise along a fault here. Exposures of the Austin chalk are good in San Pedro Park and also near Trinity University. Payaya Indians relished the springs prior to the arrival of the Spanish who first visited the site in 1709, and named it "el Agua de San Pedro." The original San Antonio village and presidio (fort) were located near San Pedro spring. The springs were set aside as a public park in 1734, the oldest in

Sinkholes in Edwards limestone in Loop 1604 roadcuts north of San Antonio

Texas, and perhaps the second oldest in the United States. Ground water flows downward toward San Antonio from the high Edwards Plateau to the north and eastward from sinks in the Nueces and Frio rivers to the west. Fractures, cracks, and cavities in the limestone blocks form a tortuous pathway for the water, which eventually surfaces at springs such as those in San Pedro Park. Notice the large cypress trees in the park—they like to have their feet in water—evidence for the continuous activity of the springs and drainageway here. Cypress trees grow along the San Antonio riverwalk for the same wet-footed reason.

Most historic San Antonio buildings are built of native stone. The most famous, the Alamo, originally a Spanish mission (Mission San Antonio de Valero) constructed at its present site in 1718, is walled with Cretaceous Austin limestone, which probably came from the quarries in Brackenridge Park. The Missions Concepcion and San Jose have walls of calcareous tufa, a soft limy rock chemically deposited in nearby springs. Missions Espada and San Juan, being farther south in the Tertiary sandstone and shale belt, have walls of Eocene Wilcox sandstone and pebbly gravel cemented by caliche. Historic buildings commonly tell the story of the local geology in their walls, because pioneering folks who built them just didn't go far afield to get building stone if they could find it nearby.

North of San Antonio are excellent new roadcuts along Highway 1604. Along this stretch of road, Edwards limestone beds are exposed, and several sinkholes can be seen where normally flat-bedded limestone beds have collapsed into holes. Surface water enters the subterranean cavern system via these sinkholes.

Travelling north from San Antonio on Interstate 10, the highway passes a large limestone quarry operation where Edwards limestone is mined for aggregate, building stone, and cement. From here the edge of the Edwards Plateau and Balcones escarpment is very evident. I-10 continues to climb the edge of the plateau toward the town of Boerne. The remainder of I-10 north of San Antonio is discussed in the Central Section.

Interstate 35
San Antonio—Laredo
147 miles

Interstate 35 between San Antonio and Laredo is a long, dry stretch of highway. The underlying bedrock along the entire length of road is Eocene sandstone and mudstone that formed one phase of Gulf-margin sedimentation about 50 million years ago. Near San Antonio the highway crosses a series of Eocene sand ridges, which stand up prominently to erosion because they are a bit harder than the adjacent mudstone sections. The band of Eocene deposits swings southward between the towns of Devine and Pearsall; for the remainder of the distance to Laredo, the road runs along the middle Eocene Claiborne band.

Near the I-410 and I-35 crossover in San Antonio, watch for sand and gravel pits, where construction gravels are mined from the coarse sediment of the San Antonio River system. Low ridges of sandy hills south of the I-410 crossover are lower Eocene in age. Coals are mined nearby from this group of rocks. Low, rolling topography forms where the alternating sandstone and Wilcox mudstone beds are eroded at different rates. The sands stand high; the mudstones are in the bottomlands. There is also a nice limestone exposure near the Shepherd Road exit.

About 15 miles south of the I-410 crossover, I-35 climbs a ridge of reddish sand. This is Eocene Carrizo sandstone, which characteristically forms red sandy hills and ridges in a distinctive band across a large part of the state. Southward, the Carrizo sandstone extends downward into the subsurface to form an aquifer which waters the Winter Garden district to the south. This exposed sand ridge, then, is an important recharge area where rainwater enters the Carrizo system. It is an obviously iron-rich sandstone. The iron comes from an iron-rich mineral called glauconite, which forms as tiny pellets, sometimes as animal fecal pellets on shallow marine shelf areas. The iron is so rich in rocks of this type in northeast Texas that the iron was once commercially mined.

Sandy red roadsides and fields continue for a few miles to the town of Moore, where a younger set of Eocene sands is encountered, as evidenced by the color change to tan. Some white caliche can be seen on hilltop ledges around here.

From Devine to Laredo, middle Eocene Claiborne sandstones and mudstones are seen sporadically in roadcuts. Oil fields and their

Cactus and thornbrush country 30 miles northeast of Laredo on Interstate 35.

nodding pumps dot the landscape, and drill rigs indicate the area is still being explored for new oil. North of Cotulla, the fields produce mostly oil, while south of Cotulla, gas is the main product. The Pearsall Oil Field near the town of Pearsall was discovered in 1936 and is expected to produce about 45 million barrels of oil from fractured upper Cretaceous Austin and Buda chalks at a depth of about 5,300 feet. Nearby Big Foot Field, north of Pearsall, produces oil from lower Tertiary delta sandstones. Stuart City and Encinal fields near the town of Encinal produce gas from reservoirs in Cretaceous limestone and lower Tertiary sandstone.

South of Cotulla, the highway crosses the Nueces River, which heads in the limestone uplands of the western Hill Country, and empties into the Gulf of Mexico at Nueces Bay near Corpus Christi. Note the chert-rich limestone gravels here which streams have eroded from the Edwards Plateau.

Thornbrush and cactus become prominent in the countryside approaching Laredo. These hardy survivors are well adapted to a climate where evaporation is very high in the searing desert sun, and where rainfall averages only 20 inches per year.

Southward from the U.S. 83 junction (about 20 miles north of Laredo) note the gravels on high hills and drainage divides amongst the generally low, undulating topography. The gravels were laid down by ancient streams that flowed on top of those high surfaces, when stream level and sea level were higher than they are today. Since the

Pleistocene warm period of 135,000 years ago, streams cut downward to adjust to lowered sea level during the cold period near the end of the Ice Age.

Interstate 37
Corpus Christi—San Antonio
146 miles

Interstate 37 heads north out of Corpus Christi, gradually climbing the river bluffs on the west bank of the Nueces River. The road follows Nueces Bay for a few miles, still running along the top of the terraces adjacent to the river. A few miles west of Corpus Christi watch for quarry faces in the Pleistocene clay beds next to the freeway, between the road and the river. The road continues to ride the drainage divide on the edge of the Nueces River valley, then it descends into the valley and crosses the river at Nueces River Park. The road travels the low, flat river floodplain before climbing the other bank up onto the Pleistocene coastal plain surface, where rich black soil and farms dominate the nearly treeless landscape for many miles. Occasional oil field pumps tell of additional black riches beneath the surface. Northeast-southwest bands of subsurface reservoirs have been tapped in this area. Oligocene barrier and strandplain sandstones of the Frio formation provide the main producing reservoirs around Corpus Christi. Northwest of Corpus Christi, fluvial and deltaic sandstones of the Frio formation are the principal oil and gas reservoirs, while

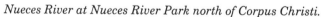

Nueces River at Nueces River Park north of Corpus Christi.

Eocene Jackson and Yegua sands form the reservoirs about halfway between Corpus Christi and San Antonio. Closer to San Antonio the fields are found in older Tertiary sandstones and upper Cretaceous chalks and limestones.

Near the town of Mathis is Lake Corpus Christi, and a State Recreation Area; the lake is a dammed up part of the Nueces River. Note the topography around the lake is suddenly rolling, not flat—the edge of the Pleistocene coastal plain extends from Corpus Christi to Mathis, and the road is now on older Tertiary sandstones, which are a bit more resistant than clay beds, so they stand up a bit more to erosion, forming topographic relief. Look for sand quarries around here, in combination with nice views to the west of Lake Corpus Christi. The white sand seen on hillcuts and quarries from 20 to 30 miles northwest of Corpus Christi is the Pliocene Goliad sandstone. It is a noted ridge-former that parallels the coastal plain and can be seen holding up ridges in a long band across much of southern Texas.

About 30 miles north of Corpus Christi, near the George West exit, the countryside flattens out somewhat, but the relief is still low and rolling, so you know you're in an area of Tertiary outcrops. The Miocene–Oligocene-age sandstone and mudstone beds underlying this area are just not as resistant as the higher standing Goliad sandstone seen previously..

Near the town of Three Rivers is a sand quarry west of the highway. Light tan to white sandstone, still of Miocene–Oligocene age is extracted for construction use. The road follows the Atascosa River valley, a tributary of the Nueces River. The headwaters of the Atascosa begin in the hills southwest of San Antonio and south of the Medina River.

Sixty miles from Corpus Christi, Interstate 37 crosses San Cristobal Creek, and quarries and cuts in whitish tuffaceous sandstones mark the northern edge of the Miocene–Oligocene band of Tertiary rocks. The abundant tuff fragments in this Catahoula sandstone tell us of Oligocene volcanic eruptions hundreds of miles to the west and northwest. North of here flatter terrain lies on marly beds of Eocene (Jackson) age, and not much in the way of rock exposures are seen.

Brush country, a few erosional divides, and gray mudstones in grassy roadcuts are seen between the Almos Creek crossing and Pleasanton. Large earthen dumps, now vegetated, are from former uranium mines in the Eocene (Jackson group) rocks near here.

A few miles north of the U.S. 281 and I-35 junction, reddish, sandy soils indicate iron-rich sandstones lie just beneath the surface. These sands are Eocene in age (Claiborne group).

94

Coming into San Antonio, watch for even redder soils and sand in grassy roadcuts just north of the Bexar County line. This bright red unit is the Carrizo sand, which forms a distinctive red band across the south half of the state.

The urban corridor south of San Antonio is now in view, and just west of I-37 and Loop 1604 are clay pits dug in uppermost Wilcox claystones, just below the Carrizo sand. East of the freeway, the town of Elmendorf hosted a large brick industry for a time, and some bricks are still being made today. Not many rocks are to be seen now until you stand and study the historic limestone walls of the Alamo!

U.S. 59
Victoria—Goliad—Beeville—I-35
72 miles

West of the historical town of Victoria, Highway 59 passes over the Guadalupe River, whose headwaters are found in the Hill Country northwest of San Antonio. Most of the road between Victoria and Goliad is on flat coastal plain terrain, until a few miles east of Goliad where the road climbs onto the upland margin of the coastal plain and the Goliad sandstone (Pliocene) forms a low ridge. The Goliad sandstone was named, of course, for exposures around the town of Goliad and several good roadcuts and locations here display this gray, cemented, pebbly sandstone. Look for the roadcut near the railroad overpass just south of town on U.S. 183, which takes you to Goliad State Park, where Mission Espiritu Santo and Presidio La Bahia are located.

Mission Espiritu Santo at Goliad.

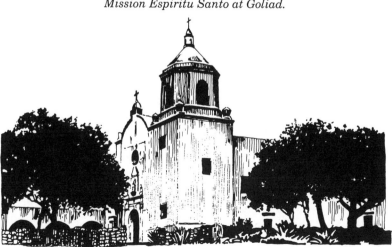

The mission was established first on the coast at Lavaca Bay in 1722, relocated to a site north of Victoria, then finally moved to its present site on the San Antonio River in 1749. The stone walls around the beautifully restored church are Goliad sandstone, quarried from the local stone near the mission. Presidio La Bahia was the site of the Goliad Massacre so dear to Texas history, where Colonel James Fannin and 351 of his men were executed by Mexican troops only a few weeks after the ill-fated Alamo fight. The battle cry at San Jacinto, where Sam Houston finally defeated Santa Ana was, "Remember Goliad! Remember the Alamo!"

From the San Antonio River bridge west of Goliad, the topographic relief produced by the resistant Goliad sandstone is noticeable. Between Goliad, Beeville, and I-35, the underlying bedrock is all Goliad sandstone. The only exception is a low area encountered a few miles east of I-35 where softer Miocene rocks are carved by the Atascosa River.

U.S. 77-83
Brownsville—Corpus Christi
152 miles

The Rio Grande is not so "grande" at the border crossing between Mexico and the United States. In fact, it is little more than an entrenched, channelized ditch here, in contrast to its broad expanse and lush green floodplain near El Paso. But considering the amount of Rio Grande water that is stored upstream behind dams and in irrigation channels, it is a wonder there is any water at all in the Rio Grande by the time it reaches Brownsville. In and around Brownsville are a number of long, narrow, meandering lakes, called "resacas," which are the oxbow lakes and abandoned channels of the Rio Grande. These are distinctive features of the Rio Grande delta which occupies the immense area of flat land bordering the river in the vicinity of Harlingen and Brownsville.

U.S. Highway 77-83 between Brownsville and Corpus Christi crosses three segments of landscape, with an urban corridor anchoring the southern end at Brownsville/Harlingen and the northern end at Corpus Christi. Flat expanses of farmland on rich brown soils, overlying floodplain clay and sand, spread for miles on either side of the highway south of Corpus Christi and north of Brownsville. These two segments form classical coastal plain terrain. In the center

Rio Grande at International Crossing in Brownsville. Mexico is to the left and the United States is to the right.

segment of the road, between the towns of Raymondville and Riviera, the roadside scenery is markedly different. Here are dry sandy soils, scrub trees, lots of cactus, and sparse grass. This region is a windswept band where sand has been blown inland from sandy shores for thousands of years. Flat sand sheets cover the ground, nearly everywhere, punctuated infrequently by sand dunes that rise above the surrounding surface. The dunes are mostly vegetation-covered and stabilized, though a bright patch of sand can be seen here and there on the flanks of a dune where the wind has somehow wormed its way through the plant cover to form a blow-out.

Another interesting landform in this desert landscape is the dry lake. Look for large patches of low ground, clay-grey in color, where plants don't seem to want to grow. This is a lake—when there's enough rain water to fill it. Distinctive tussocks of grass around the edges of these dry pans apparently get their feet wet frequently enough to hang on for dear life at the water's edge.

The scrub tree and grass region of blowing sand is mostly on the extensive property of the world-famous King Ranch, where Santa Gertrudis cattle were developed into a distinctive North American breed.

97

PADRE ISLAND NATIONAL SEASHORE

Access to Padre Island National Seashore is from Corpus Christi via Texas 358, a divided highway that splits off Interstate 37 and heads southeast through the city. The road first crosses a narrow bay, called Cayo del Oso, then passes over the Encinal Peninsula on which the Corpus Christi Naval Air Station and Padre Island National Park Headquarters are located. The peninsula is a preserved Pleistocene barrier-bar, built during an interglacial period when sea level was higher than it is today. Flour Bluff is the highest point on the peninsula, and of the landscape for twenty miles inland. The highway becomes Park Road 22 as it arches over the wide expanse of the lagoon, Laguna Madre, on the JFK Causeway. From the causeway's high vantage point you can see Laguna Madre extending to the horizon to the south, and get a feel for the shape and size of Padre Island. Note also the straight line of the intracoastal waterway and its adjacent spoil piles.

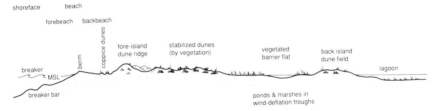

Profile of environments on North Padre Island.

A vast array of goods and materials move up and down the Texas coast in barges plying the protected waters of the lagoons via the intracoastal waterway. The waterway segment near Padre Island was dredged in 1949 and is maintained at 12 feet deep and 125 feet wide by the U.S. Army Corps of Engineers.

Follow Park Road 22 to enter the National Seashore. The two-lane paved highway winds southward down the center of the island, where it ends as a beach access road about five miles within the park. Southward beyond this point, car travel is on unpaved beach road, and four-wheel drive vehicles are required for extended travel down the island.

Padre Island is a wonderland of natural geologic processes. Wander down the beach and watch the waves move sand on the beach. See longshore drift in action. Look for layers of dark heavy minerals, which are moved at a different rate than the lighter quartz sand grains. Note how shells change in size and type along the forebeach.

*View of dunes looking toward vegetated flats and lagoon.
Note the wind ripples.*

Watch for ghost crab holes as you approach the edge of the fore-island
dunes. And, while in the dunes, see if you can spot wind-eroded dune
edges where large cross-strata are exposed to show the internal
structure of the dunes. Observe how the dunes are controlled by the
hardy vegetation—sea oats, morning glory, sea purslane, beach tea,
and panicum, to name the most common. In a few places the wind has
broken sand free from the plants' tight grip to form free-standing sand
dunes that take on their own shape and identity. On the landward side
of the dunes, you can see how the dunes decrease in height and grade

Dunes and internal bedding.

into a vegetated barrier flat, where tall grasses and low shrubs cover a surface that is not infrequently inundated by water as storms and high tides churn over and around the barrier island. Ponds and small lakes occupy low spots on the flats.

Another set of dunes is to be found on the edge of the lagoon where sand is blown from the free-sand edges of the lagoon's inner shore. Windswept flats around the lagoon are the product of tidal variations of water levels in the lagoons. Tides are spawned both by the moon's pull and by wind surges that pile up water in the lagoons, only to subside, leaving behind the flat-exposed surface. Algal mats anchor the tidal flats at many locales, appearing as dark areas amongst the lighter sand. You might also be lucky enough to find gypsum rosettes in sediments of the tidal flats. These crystals of calcium sulfate form as highly saline water seeps into the sediments, then dries, leaving behind the gypsum as a precipitate. With successive wetting and drying cycles, the crystals grow into rose-shaped clusters.

Padre Island offers a perfect natural laboratory in which one can learn about coastal environments, but also observe the processes and products of wind, water, tides, hurricanes, and waves in dynamic action.

If you would like to know more about Padre Island's environments and processes, an excellent, beautifully illustrated publication is recommended: *Padre Island National Seashore—A Guide to the*

Wind deflation. Wind blew from right to left; small pebbles acted as wind break protecting a downwind tail of sand from blowing away.

Sea oats on a dune. Wind has excavated the dune revealing inclined, internal structure of the dune.

Geology, Natural Environments and History of a Texas Barrier Island, by Bonnie R. Weise and William A. White, published in 1980 by the Bureau of Economic Geology, the University of Texas, Austin 78712. Look for the book at the National Park Service Visitors Center.

ROCKPORT/FULTON AND ARANSAS NATIONAL WILDLIFE REFUGE

Located about 35 miles north of Corpus Christi on Texas 35, the small beach towns of Rockport and Fulton are the ports for tour boat cruises of the Aransas Bay area. The focus of these cruises is to view the large populations of winter birds that descend yearly on Aransas National Wildlife Refuge. The main attraction is the exquisite whooping crane, which is gradually fighting its way back from the brink of extinction.

The coastal geology of the Aransas Bay area is responsible for the habitat which attracts these birds. As shown on the map, the Aransas Bay area is dotted with numerous shallow bays, only a few feet deep, which once were river mouths that have been inundated by rising sea level since the last ice age. The bays are rich in marine life, especially oysters, which grow in these estuarine waters fed by the nutrient-laden streams. Many places in the bays are nearly choked by oyster

101

reefs—many more reefs occur here than appear on the map. Many low islands are dredge spoils from the construction and maintenance of the intracoastal waterway.

The barrier islands, St. Joseph and Matagorda islands, protect the inner bays from wave pounding in the open surf, so the bay environment is calm compared to the barrier coast. But, it is probably the wide expanse of mud flats and their rich marshes on the inside of the barrier islands that most attracts the birds. The muddy and sandy flats are the depositional products of storm washovers that periodically spill over the barriers. The Aransas Bay area has the widest, most extensive barrier washover complexes on the entire Texas coast. Though the tidal range is only a few feet, water in shallow tidal

Environments in Rockport/Fulton and Aransas Bay area.

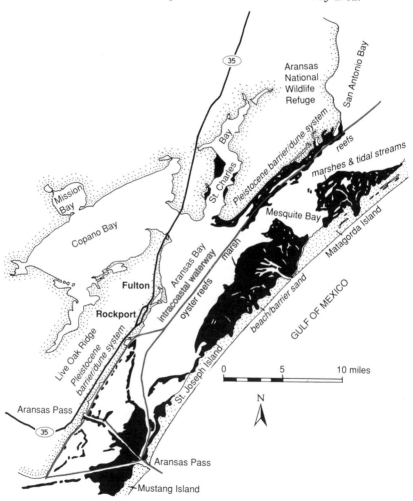

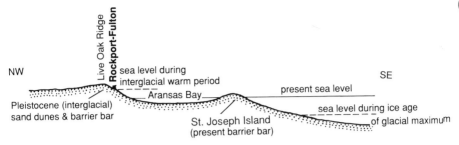

NW

Live Oak Ridge

Rockport-Fulton

sea level during
interglacial warm period

present sea level

SE

Aransas Bay

Pleistocene (interglacial)
sand dunes & barrier bar

sea level during ice age

St. Joseph Island
(present barrier bar)

of glacial maximum

*Cross section showing fluctuating sea level and associated landforms
around Rockport and Fulton.*

channels flushes through the marshes twice daily, supplying nutrients to marsh and bird alike.

The combination of all these geologic sub-environments in one place is what attracts the thousands of water fowl and shore birds, including whooping cranes, to the Aransas Bay area. So far the challenge of habitat protection, balanced with human endeavors of boating, shipping, recreation, oil production, and commercial fishing has been met, though probably not to the satisfaction of everyone.

Before leaving the Rockport area, note the hill, called Live Oak Ridge, that parallels the coast in and around the town of Rockport. The ridge is a line of stabilized sand dunes, which were once part of the Pleistocene (Ice Age) barrier island system, now covered by vegetation and notable live oak trees. The live oaks are shaped and bent land-ward by the prevailing onshore wind which continually blows in from the Gulf.

*Port Aransas. Jack up drill rig (left) and a production tower (right)
ready to be towed out to sea to, respectively, drill and produce oil.*

U.S. 83
Laredo—Harlingen
165 miles

The highway follows the Rio Grande valley from Laredo to Harlingen, but the river is rarely in view, except around the town of Zapata. There the road crosses fingers of Falcon Reservoir, which is a dammed lake on the Rio Grande.

As the road makes its southward bend in Laredo, note the exposures of tan sandstones in high cut banks of the creek. These are beds of the Laredo formation (Eocene age), a sandstone and claystone unit originally deposited in the shallow sea margin of the early Gulf of Mexico. The road follows the Laredo beds from Laredo to Zapata near Falcon Reservoir.

The countryside is sandy, dry—only 18 to 20 inches of rain fall per year here—aptly called the Arid Plains on physiographic maps. The dryness results in a pleasant climate year round, and consequently,

Channels in Laredo sandstone on US 83, 25 miles south of Laredo, Texas.

Cross-bedding in Laredo sandstone on US 83, 25 miles south of Laredo, Texas.

the Rio Grande valley is attracting more and more "Snow Birds" and "Winter Texans" each year, who seek relief from colder northern climates.

About ten miles south of Laredo, the sharp mountain profile of the Sierra Madre Oriental can be seen on the Mexican skyline to the west. Watch too for patches of badlands erosion where watery rills have etched headward into banks of local drainages after thunderstorm downpours.

About 25 miles south of Laredo, the road ascends a knoll of Laredo sandstone, where an excellent roadcut displays sandy channels and cross-bedding in the sands.

A scenic view of the Rio Grande and Falcon Reservoir shows up on a high road point north of Zapata. The highway crosses the upper reaches of the Falcon Reservoir near Zapata. Falcon Dam is operated jointly by the United States and Mexico, under supervision of the International Boundary and Water Commission. Falcon Reservoir attracts boaters and bass fisherman, and provides irrigation water and power to the region.

Goliad sandstone on mesa near Garceno and Villareales about 12 miles northwest of Sullivan City on US 83.

U.S. 83 swings southward at Zapata, and follows the next younger Tertiary unit (Yegua sandstone) for about 15 miles. A wide variety of crops are grown on the flat, fertile, brown-soil fields situated on the Rio Grande floodplain. Mild weather combined with irrigation water from the Rio Grande make the lower Rio Grande valley one of the richest produce regions in the nation.

Eocene sandstones and claystones continue to form the surrounding bedrock between Falcon Dam turnoff and the town of Garceno. Between Rio Grande City and Sullivan City, ridges and mesas are quite prominent on the side of the highway away from the river. A hard, pebbly gray sandstone holds up the top of the mesas above pinkish claystones and siltstones. Quarries and natural exposures are fairly common along this stretch. The ridge-forming sand is the Pliocene Goliad sandstone, named for the town of Goliad, which is northeast of Corpus Christi. The Goliad sandstone forms a nearly continuous ridge from here to Goliad. The road travels from Rio Grande City nearly to McAllen, either along the edge of the Goliad ridge or on top of it, before dropping down to the flat, level surface of the coastal plain just north of McAllen. The terrain is flatter, grassier, moister, soils are browner and citrus groves become the modern day forest surrounding the highway. Trailer camps proliferate too, and Mission, Texas usually welcomes winter visitors with signs and banners saying "Winter Texans Welcome at Mission, Texas." Needless to say these visitors add significantly to the economy of the Rio Grande valley.

From McAllen to Harlingen, strip city forms the scenery on either side of the road.

Texas 100, 48
Port Isabel and South Padre Island

From the junction of U.S. 77 - 83, Texas 100 heads eastward across the flat floodplain of the Rio Grande. The soils are rich, brown, and support extensive agriculture. At Laguna Vista, the road turns right and you get the first view of Laguna Madre, the lagoon, that lies between the coast and barrier island. The lagoon is noticeably calm, because the water is only a few feet deep, so big waves can't build up. In the ocean, the wind presses on the water surface, creating large waves which are really circular motions of water that extend many feet downward from the surface. In a shallow lagoon, large waves don't form because the water is too shallow to allow the development of the deep circular motion inherent in big waves. Instead, the wind produces lots of choppy little waves in shallow water lagoons. Notice there is very little beach along the lagoon. Since big waves don't churn the bottom much, sand simply isn't piled along the lagoon's edge in most places. The road parallels the lagoon into the coastal town of Port

Port Isabel lighthouse. Erected in 1852.

Isabel, where the old Port Isabel Lighthouse still stands as a monument to 19th Century coastal trade and commerce. The lighthouse was erected by the U.S. government in 1852, extinguished during the Civil War, then discontinued from 1888 - 1895. The light was permanently extinguished in 1905, but is now maintained as a Texas State Historical Structure by the Texas Parks and Wildlife Department.

The causeway between Port Isabel and South Padre Island gives a high vantage point to see the modern coastal environments of lagoon, tidal flats, sand dunes, and barrier island. Note especially how narrow the South Padre barrier island is compared to North Padre Island near Corpus Christi. Longshore drift of sand along this southern part of the Texas Gulf Coast moves northward, driven by the predominant southeast to northwest winds. The major sand source is the Rio Grande, augmented by the smaller Texas rivers. Sand drifting longshore from the north and from the south converge near Corpus Christi to build a wide barrier island system there. Hence, the North Padre Barrier Island is wider than South Padre Island.

South Padre Island. Partly vegetated dunes behind beach.

Drive north from the beach town of South Padre Island to see a marvelous stretch of virtually untouched beach, dune, and tidal flat complex. The development on South Padre Island has (so far) been confined to the very southern tip of the island, leaving plenty of open space along a ten mile stretch north of town.

It is easy to spend hours amongst the dunes looking at their internal cross-bedded structures, seeing how sand moves across the dune's steep faces in low ripples, or observing myriad insect tracks that dash across the dry sand. Vegetation plays some role in trapping sand in dunes, though the dunes here are not stabilized by plants nor even extensively covered by the vegetation. Lots of free sand blows around.

Note how the high edge of the dunes is along the beach/Gulf side and how the dunes trail off into the lagoon on the other side. The source of the sand is therefore the beach and onshore winds tend to pile up the sand directly behind the beach.

The tidal flats on the lagoon side of the barrier vary in extent. But the wide expanse of tidal flats in this area is somewhat amazing considering the tidal range in the Gulf of Mexico is very low — only one or two feet. A lot of very low-relief land is inundated, then dried, as the low tidal waters sweep in and out across the flats twice daily.

The best tidal flats are seen a few miles south of Port Isabel along Texas 48. Sand dunes are also abundant and easily discernible along this stretch of road. In fact, the first ten mile stretch south from Port Isabel on Texas 48 crosses a delightful area of near-wilderness where coastal environments in virtually their natural state can be viewed and enjoyed from the highway. The vastness of Bahia Grande—aptly named as the Great Tidal Flat—is impressive.

The dunes surrounding the Bahia stand up clearly, their long linear shapes forming ridges above the flat terrain. Grass flats, bays, round lagoons, and lakes complete the environmental scene. Note the tall Yuccas on sandy slopes of dune ridges. It is dry here—only about 20 inches of rain fall annually in this southernmost corner of Texas. This is a nice, quiet get-away place!

Texas 48 continues into Brownsville, the evidence of habitation increasing incrementally nearer to town. The road follows the inter-coastal waterway which connects Brownsville with sea traffic in the Gulf. Fishing boats stand above the dry surrounding landscape, ten miles from Port Isabel, looking oddly out-of-place this far from the Gulf.

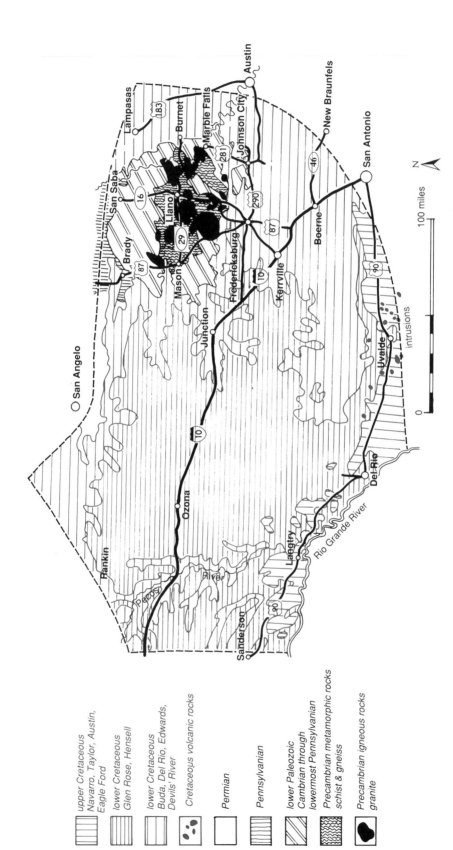

Geologic map of Central Texas.

upper Cretaceous
Navarro, Taylor, Austin,
Eagle Ford

lower Cretaceous
Glen Rose, Hensell

lower Cretaceous
Buda, Del Rio, Edwards,
Devils' River

Cretaceous volcanic rocks

Permian

Pennsylvanian

lower Paleozoic
Cambrian through
lowermost Pennsylvanian

Precambrian metamorphic rocks
schist & gneiss

Precambrian igneous rocks
granite

III
Central Texas
Hill Country, Caves, and Plateaus

Central Texas is a wonderful recreational area, where millions of people annually enjoy the numerous canyons, lakes, streams, caverns, springs and rocks in this region. The bountiful scenery is principally due to the diverse geology of the area and the rocky origin tale is a complex but delightful one. Once you know the story, your enjoyment of the Edwards Plateau and Hill Country scenery is enhanced even more.

The saga unfolds about ten million years ago, though the ultimate geologic chapter was first written long, long before that. Today Central Texas stands about 2,000 feet above sea level. The uplifted Cretaceous rocks are marine sandstones, limestones, shales, and dolomites which were originally deposited in the ocean below sea level. This, of course, means the entire Edwards Plateau has been physically elevated as a mass about 2,000 feet. Many geologists believe this uplift occurred about ten to twenty million years ago during Miocene time, as part of a regional upwarping that occurred across the western United States. Other geologists argue that Central Texas was elevated long before the Miocene, perhaps in early Tertiary time, and the Edwards Plateau was left standing high as movement along the Balcones fault in Miocene time caused the area on the southeast side of the fault to drop. Whether we believe a one-phase or two-phase structural sequence for Central Texas, the thick layers of Cretaceous sedimentary rocks have been elevated 2,000 feet essentially undeformed. The uplift did not seem to fold or contort the rocks much at all, because they stand remarkably flat-lying today, much as they were when originally deposited in the ocean 100 million years ago during Cretaceous time.

The result was a high-standing, flat-surfaced limestone bench which is now bordered by a steep, southeastern fault face, known as the Balcones escarpment. (Balcones is Spanish for "balcony", which perfectly describes the escarpment.) The escarpment is a fault-line scarp, caused by faster erosion of the softer rocks which lie to the south and east of the Balcones fault.

Over the last ten million years erosion by streams has attacked the margins of the plateau, and the streams continually cut headward as well as downward in an erosional frenzy to flatten this highland. And, in the span of a mere ten million years, the job is about half completed. The west half of central Texas remains a high flat plateau, while the east half, called the Hill Country, is deeply eroded. This duality of topography certainly contributes to the geologic variety found in Central Texas. The interaction of water with limestone in the Edwards

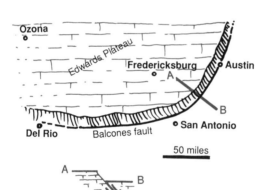

Ten million years ago the Edwards Plateau was uplifted along the Balcones fault.

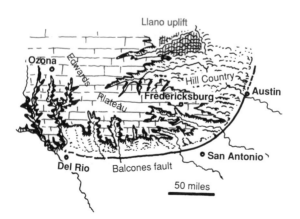

Since then stream erosion has carved into the Edwards Plateau, exposed the Llano uplift, and created the Hill Country.

Plateau weaves an intriguing tale of groundwater, limestone dissolution, and beckoning subterranean caves, told in the following section entitled "Texas Caverns." The unveiling of an archaic mountain range through deep erosion in the adjacent Hill Country is explained below under the heading "Llano Country." Read on to learn more about the story behind the geology of Central Texas, and travel on to enjoy the geology first hand.

Texas Caverns

In the Texas Hill Country and Edwards Plateau region, limestone is everywhere. Thousands of square miles of Cretaceous limestone are exposed in stream banks, hillsides, roadcuts, and cliffs. The hard, limy rock resists surface erosion better than softer surrounding shales, and so the generally flat-lying limestone beds stand out as ridges and cliffs and form ledges on top of hills and mesas.

Natural Bridge Caverns.

1. Percolating ground water enlarges fractures in limestone as it seeks its way down ward toward the water table. 2. Passageways enlarge and the subterranean cavern system moves water laterally at water table level toward streams. Streams continue down-cutting. Springs emerge from banks at or above the water table level. 3. Streams cut even deeper. Water table lowers, new, deeper passageways are created. Higher caverns are left dry, except for minor dripping water that builds dripstone formations.

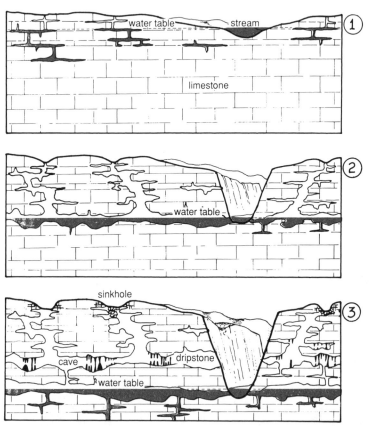

But as hard and indestructible as limestone appears at the surface, underground is an almost opposite scene. For beneath the surface of the Edwards Plateau, the limestone is riddled with holes and cracks, endless caverns, and intricate subterranean passageways. Limestone, you see, is soluble. Seemingly gentle rain drops pattering on a flat limestone outcrop are not gentle on the limestone at all. Rainwater contains dissolved carbon dioxide from its passage through the air, which makes even natural rain slightly acidic. (The slightly acidic rainfall is normal, therefore, even in the absence of pollutants that cause "acid rain," which we read about in the press these days.

116

Pollutants add even more carbon dioxide, or sulfur dioxide, or nitrous oxide to the water in the air to create worrisome levels of carbonic, sulfuric, and nitric acids, respectively, in rainfall in some areas.)

For millions of years, this weak acid has percolated downward through tiny fractures and along openings between bedding planes in the subterranean limestone of the Edwards Plateau. The groundwater slowly dissolves the limestone along these microscopic passageways, and bit by bit the passage-ways are enlarged. The water percolates downward seeking a level, which hydrologists call the water table, then moves laterally toward a nearby surface river, always dissolving the limestone. Passageways are enlarged until a subterranean channel is created at water table level, where ever more groundwater flows laterally to streamside. A cave has thus been created.

Meanwhile, all has not been quiet at the surface either, for rivers and streams have been doing their job of cutting downward. As the rivers reach a new lower level, an equivalent level is sought in the subterranean system, and a new, lower, main passageway is created, leaving the upper, former main passage dry. But, not quite dry, because rainwater continues to fall on the plateau above, and groundwater continues to percolate downward through main passageways. But, some water still leaks downward through tiny fractures to drip from the ceiling of the abandoned, upper-level caves. This water is the same as always; it continues to be slightly acidic, it continues to dissolve limestone from the walls of fractures, and to carry away the dissolved calcium in a watery solution. But, the story now becomes intriguing, because a little chemistry trick occurs just as the water droplet emerges from the tiny crack in the cave ceiling to hang motionless for a moment before dropping to the ground floor far below. In the crack above the ceiling, the drop is carrying a full load of calcium

Stalactites form where calcium-charged water drips from a local low spot on the cave ceiling. Stalagmites build up from the spot of water-impact on the cave floor.

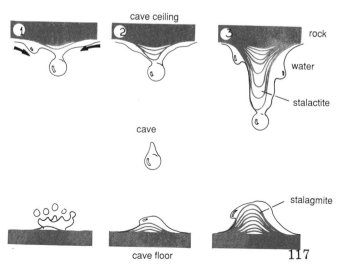

as well as a balanced load of carbonate (carbon dioxide) from the dissolving action on the limestone (which is calcium-carbonate). As the drop hits the crack opening in the cave ceiling, it is carrying more carbon dioxide than is in the cave air, so a bit of carbon dioxide is released to the cave air, to even things up, chemically speaking. As soon as that happens, the remaining water in the hanging drop becomes instantly supersaturated with calcium and just before the drop lets go, the extra calcium (along with some carbon dioxide) is left behind on the ceiling as a thin, precipitated ring or film of pure calcium-carbonate (the mineral calcite), and a stalactite is born.

Falling through black cave space, the drop loses more carbon dioxide to the air, and by the time it splashes on the floor, it is again supersaturated with calcium. Another bit of calcite is left on the cave floor at the point of the droplet's impact, and a stalagmite has started to form. Repeating the process drop by drop for countless millenia creates the wonderful, artful and awe-inspiring dripstone formations we love to see in caves. Stalactites hang from the ceiling, stalagmites rise from the floor, and where they meet, solid columns are formed. (Stalactites with a "c" hang from the ceiling; stalagmites with a "g" grow from the ground.) Delicate soda straws (hollow, tubular stalactites) form when one drop at a time leaves the ceiling from a tiny hole, leaving a perfect calcite ring behind; successive rings simply grow into a tube. Droplets emerging from side walls deposit calcite sideways, upwards, and downwards in irregular patterns to form odd-shaped twigs, fans, "butterflies", and "fishtails"—all called helictites. Some speleologists (cave scientists) have even suggested air currents may

Forms of dripstone.

sinkhole

limestone

stalactite

drip curtain

soda straws

helictites

cave pool

collapse blocks

column

stalagmite

dome

travertine terrace

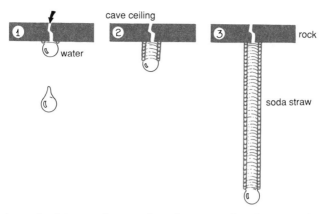

Perfect rings of calcite are deposited on the cave ceiling by water drops emerging from a tiny crack or hole. Over time, the rings grow to hollow tubes called soda straws.

blow droplets awry, to explain the many angles at which helictites grow. Water emerging in a continuous line along a crack lays down a line of calcite, and continued growth results in the marvelous drip curtains seen in many, but not all, caves. Though pure calcite is white, some cave dripstone is colored yellow, tan, brown and even red. These warm colors are derived from varying amounts of iron carried in groundwater and precipitated along with the calcium.

Large passageways frequently are enlarged to caves as unstable roofs collapse, creating sinkholes above and piles of collapse debris below at the cave floor. Sinkholes are common over much of the Edwards Plateau and are significant openings where rainwater enters the underground piping system to form the aquifers that provide drinking water for much of the Hill Country and the San Antonio area. The Edwards aquifer is consequently carefully watched and managed by the citizens of this region.

Edwards Plateau Aquifer.

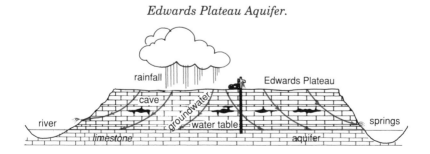

119

The story of water interacting with limestone in the Edwards Plateau/Hill Country is a fascinating one, because it ties together rainfall, groundwater, water tables, aquifers, cave formation, stalactites, springs, river levels, and erosion. It hopefully evolves into an increased awareness and understanding of this single, large, complex, dynamic, interactive, interwoven rock-water system.

Fortunately, a number of beautiful caverns in Hill Country limestone are open for public viewing (see map). *Natural Bridge Caverns*, 12 miles west of New Braunfels off Texas 46 on FM 1863, features marvelous dripstone, giant rooms, and cave pools. *Cascade Caverns*, five miles southeast of Boerne off U.S. 87, has a 90 foot waterfall as a main attraction. *Cave Without a Name*, 11 miles northeast of Boerne off FM 474, has stalactites, stalagmites, and soda straws. *Inner Space Caverns*, off I-35 one mile south of Georgetown, has creatively lighted flowstones, and the remains of Ice Age mammals. *Wonder World Cave*, Bishop Street in San Marcos, discovered in 1893, exclaims itself as an earthquake-formed cave. *Longhorn Cavern*, 11 miles southwest of Burnet on Park Road 4 off U.S. 281, is a large, near-surface cave of historical interest: it was a secret Confederate gunpowder manufacturing and storage site in the Civil War. All the other caves are carved in Cretaceous limestone; Longhorn Cavern is etched in much older Paleozoic limestone. *Caverns of Sonora*, 8 miles west of Sonora on I-10 (go south on FM 1989 for seven miles) displays spectacular, naturally-colored dripstone, along with helictites and soda straws.

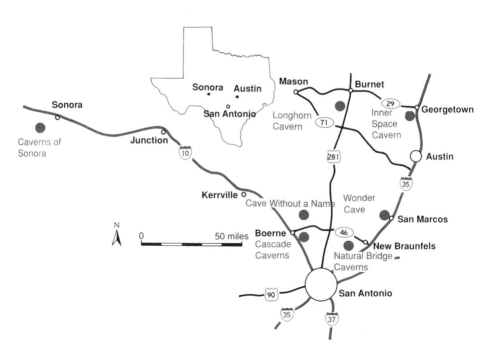

The Llano uplift is structurally high but is topographically low.

Llano Country

A unique and spectacular group of rocks is found in the center of the state, north of Fredericksburg, surrounding the town of Llano. Here, hard granite knobs reflect pink in the harsh highland sunlight, while tortured, black schists absorb the sun in dark rings around the central igneous core. This assemblage of ancient, glinting, crystalline rocks is more expected in the harsh crags of Colorado than in the soft, rolling, calcareous canyon country of central Texas. But, here they are, and the origin of the Llano uplift is a fascinating tale!

The legend begins deep in the geologic past, for the Llano rocks are ancient indeed, probably the oldest in Texas, at a venerable 1.35 billion years. Buried, squeezed, melted, faulted, then uplifted, the granite and schist form the heart of once-lofty mountain ranges which time has eroded flat, only to be later born again, raised to near surface, and finally exposed in central Texas by erosion.

Try to imagine the Precambrian scene of 1.35 billion years ago, when the area that is now central Texas lay in an ocean basin off the coast of North America. Sediments poured into this sea from adjacent mountains, building a coastal plain and continental shelf, not unlike the Texas coast today. First, a mix of rhyolitic volcanic rocks and associated tuffaceous sediments were shed from the young mountain ranges to accumulate the Valley Spring wedge of poorly sorted sediments in the sea.

As the mountain range wore down over the next few million years, a second wedge of dark, fine-grained, well-sorted muddy sediments, called the Packsaddle sediments, was laid over the top of the Valley Spring wedge. One billion years ago, in an early episode of continental drift, the edge of North America collided with another land mass. Trapped in the subduction zone between the colliding continents, the thick Packsaddle and Valley Spring sediment wedges were squeezed and folded and heated with such intensity, that all the sedimentary minerals changed—metamorphosed—into new crystalline forms. The Valley Spring gneiss and Packsaddle schist thus were born.

Not only did the rocks reform, but a mountain range rose in the melded zone of the new continent. Rocks buried even deeper beneath

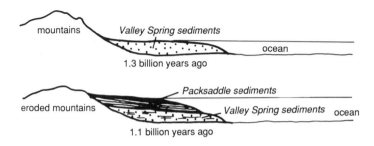

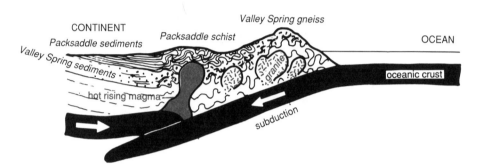

How the Llano rocks came to be formed in the Precambrian over one billion years ago.

the range in the subduction zone melted, and the ensuing magmatic gruel, lighter in weight than surrounding rocks, rose through the overlying rock column as large spherical bodies of red hot liquid. Eventually the subductive fires went out and the magma, the metamorphic rocks, and the mountain range itself cooled to surface temperatures, and the stygian magma solidified into granite.

The earth doesn't like topographic irregularities, so for the next few hundred million years it unleashed its clawing monster, Erosion, to wear the mountains down to a preferably flatter surface. By the time life was first abundantly preserved in the fossil record (600 million years ago), the old range was gone, worn to a table-top; and, along the very line where North America and its land mass neighbor once were joined, the two continents drifted apart. Ocean water lapped over the flattened margin of North America once again, and for the next few hundred million years, beds of Paleozoic limestone, sandstone, and mudstone were deposited over the remnants of the once-tall mountain range.

Then, three hundred million years ago, North America collided again with other continents to form the giant land mass, known as Pangaea.

And, expectedly, a mountain range rose across Texas along the same curved continental juncture. Geologists have named these mountains the Ouachita Range. The familiar geologic cycle was repeated once more, as erosion reduced the Ouachita Range to a flat surface. After 100 million years of quiet stability, the earth rumbled again, and Pangaea cracked and split apart, North America separated from the European continent and headed westward, while the southern margin sagged to form the early Gulf of Mexico in Jurassic time. As the Gulf continued to deepen, Cretaceous marine sediments were laid over the eroded surface, where Precambrian igneous rocks and

About 1 billion years ago, mountains rose, rocks were metamorphosed, and granitic batholiths emplaced.

In 400 million years, erosion reduced the range to a nearly flat plain, though 800' hills were left here & there.

Geologic story of the Llano region.

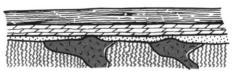

Sediments were laid over the plain during the Paleozoic.

In a burst of Mesozoic mountain building, the Llano was tilted, faulted & eroded.

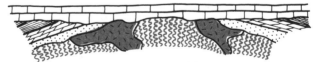

Cretaceous rocks covered the Llano as seas invaded it.

Cretaceous

Paleozoic

Precambrian

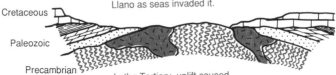

In the Tertiary, uplift caused renewed erosion; Cretaceous rocks were removed to reveal the Paleozoic & Precambrian rocks.

123

tilted Paleozoic rocks together formed the underlying beveled surface. At the end of Cretaceous time (60 million years ago), when the Rocky Mountains were rising to the west, the Gulf sagged more rapidly again along the Ouachita Mountain line, and sandy and muddy sediments poured southward and southeastward into the Gulf. For the next 50 million years Texas grew 250 miles into the Gulf on these sedimentary deposits. A structural push in the Tertiary drove the Edwards Plateau and Llano country straight upward about 2,000 feet. And, predictably, the faults along which this rise occurred follow the old Ouachita Mountain line. Erosion of the plateau began, and during the last ten million years, the thick Cretaceous sequence of rocks has been penetrated, cut into valleys and ridges, and peeled off to expose the igneous and metamorphic rocks of the Llano country.

This long and complicated geologic tale is told in the rocks of the Hill Country and Llano uplift. The story can really only be appreciated by reading it outdoors amongst the rocks.

Geology near Enchanted Rock.

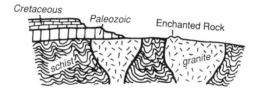

Cross section of the Llano country west of Austin.

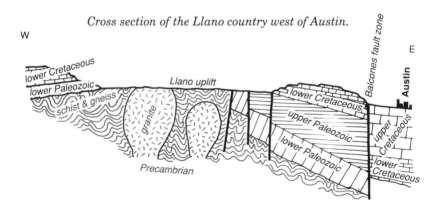

124

Interstate 10
San Antonio—Kerrville—Bakersfield
284 miles

Lots of geology can be seen along this stretch of expressway. The road passes many roadcuts and natural outcrops as it traverses the Hill Country near San Antonio and tops out on the Edwards Plateau farther west. It is mainly limestone terrain, where miles and miles of flat-lying, but uplifted Cretaceous rocks dominate the scene and handsomely illustrate the broad extent of the Cretaceous sea which covered the region. Fossils of marine snails, clams, oysters and sea urchin spines are especially abundant and a stop at virtually any roadcut will turn up fossils.

The Edwards Plateau was elevated about 2,000 feet above sea level along the Balcones fault. As you drive north out of San Antonio the highway crosses the Balcones fault system, actually a parallel series of faults. Just north of the I-410 crossover, and looking north, it is easy to see the high hills of the uplifted "balcony" ahead. Several large quarries are located near the road on the northern outskirts of San Antonio. The Edwards limestone is mined here to make cement and road gravel. The thick-bedded character and flat-lying nature of the Edwards limestone is marvelously exposed in these quarries.

Immediately north of the big quarries the road climbs uphill, and thereby crosses the main Balcones fault. Several highway cuts display the yellowish, thin-bedded limestone and marl character of the lower Cretaceous Glen Rose limestone, which I-10 follows for about 50 miles to just past Kerrville.

The area between San Antonio and Kerrville is known locally as the "Hill Country," and geologically represents the deeply dissected edge of the Edwards Plateau, where stream erosion has carved away at the

Cross section along Interstate 10 between San Antonio, Kerrville, and Bakersfield.

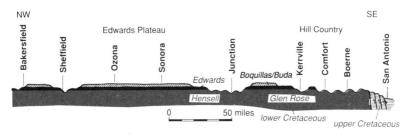

125

Excellent roadcut of
lower Cretaceous
limestone, dolomite,
and claystone along
Interstate 10 at rest
stop east of Kerrville.
Note distinct,
horizontal beds with
no folding.

Fossils at above roadcut. The arrow points to high-spiraled snail fossil. The
quarter is for scale.

uplifted sedimentary rock section for the last several million years. Watch for stair-stepped hills which are typical of Glen Rose terrain. The alternating hard limestone and soft marl beds erode unequally, so "stair-steps" are produced.

Near the town of Boerne (pronounced "Bernie") are Cascade Caverns, where stalactites, stalagmites, and a large underground waterfall are featured. The geology of limestone caverns in the Hill Country is described in the section on "Texas Caverns."

Near the town of Kerrville, the upper parts of hills are composed of light gray, thick-bedded Edwards limestone, that rests on top of the Glen Rose limestone, and which we see between here and San Antonio. The Glen Rose shows up in the lower parts of hills in the low drainage area of the Guadalupe River around Kerrville.

West of Kerrville, about at the Gillespie County line, are a series of wonderful roadcuts where broken-up limestone (breccia) is nicely shown, and where local collapse folds can be seen in the Edwards limestone. These features indicate caverns must be located below the surface. The overlying rock has collapsed into the caverns. Extensive solution cavities and caves are characteristic of the Edwards limestone.

About twelve miles west of Kerrville, the road finally tops out on the Edwards Plateau, and the surrounding flat countryside tells the tale. It is now clear that the Hill Country is the result of intensive erosion, which has removed virtually all the Edwards limestone and carved hills and valleys down into the underlying Glen Rose rocks. The interstate heads down through the Edwards into the Glen Rose again along the erosional drainage of the Llano River near the town of Junction. Excellent natural river-cut outcrops are all around Junction. The bright red beds of the Hensell sand, a sandy equivalent to the Glen Rose limestone, are particularly noticeable in the deeper valley cuts in this area.

The road follows the Llano River drainage for about twenty miles west of Junction, then climbs to Edwards Plateau level again on the way to Sonora. Watch for roadcuts on hilltops about eight miles east of Sonora where we see dark gray and brown beds of nodular marl. These are beds of the Buda formation, which overlies the Edwards limestone and caps hilltops over much of the Edwards Plateau country.

At Sonora watch for signs to Sonora Caverns, located south of town on F.M. 1989. This is a gorgeous cave, and if time permits, the caverns are a worthy side trip. Again, refer to the Texas Caverns section for a review of cave geology.

West of Sonora the change in landscape from rounded, fairly well-vegetated hills near San Antonio to steep-sloped, rocky, poorly vegetated mesas is by now a very striking comparison. The rocks are the same in both areas—flat-lying limestone—but the rainfall, hence vegetation, hence weathering, is drastically different. Average rainfall near San Antonio is 28 inches per year, but is quite desert-like at 18 inches per year west of Sonora. The junipers, low shrubs, rocks, and mesas are starting to look like real west Texas cowboy country!

The terrain is quite flat for most of the way between Sonora and Ozona, though there are several good roadcuts in Buda rocks near Ozona.

Low mesas and isolated buttes become more and more common westward. An interesting side trip on U.S. 290 takes the traveller over typical Texas mesa country, and past old Fort Lancaster, where the original camel patrol played an important role in Texas history. Both I-10 and U.S. 290 cross the Pecos River at Sheffield, and the Texas 349 —U.S. 190 loop northward through the town of Iraan passes the Yates Oil Field, one of the giant fields of North America. Discovered in 1926, contrary to the watchword "no oil west of the Pecos," the field has produced over one billion barrels of oil from cavernous Permian limestones only 1,000 feet below the surface and still produces nearly 100,000 barrels of oil per day.

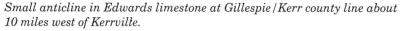

Small anticline in Edwards limestone at Gillespie / Kerr county line about 10 miles west of Kerrville.

Lower Cretaceous Edwards limestone in Interstate 10 roadcut at Kerrville.

West of Sheffield, I-10 follows the old Spanish trail along the bottomlands of Four Mile Draw. Hillside exposures of Edwards limestone, capped by Buda formation surround the roadway.

Note how the bottomlands are becoming broader and the mesas more widespread as the expressway heads on into Bakersfield.

The I-10 extension from Bakersfield westward to Ft. Stockton is described in the West Texas section.

U.S. 87
Comfort—Fredericksburg
23 miles

From the town of Comfort, U.S. 87 heads north toward Fredericksburg, passing low hills of Glen Rose limestone for several miles as the road climbs away from the Guadalupe River drainage. In a few miles the road has reached the Edwards limestone where the surrounding plateau countryside is noticeably flat. The road continues on the plateau for about seven miles before again descending through Glen Rose beds as the topography cut by the Pedernales River drainage is encountered. The road crosses the Pedernales River a few miles south of Fredericksburg where erosion has cut into the Hensell sand, which replaces the Glen Rose limestone to the west.

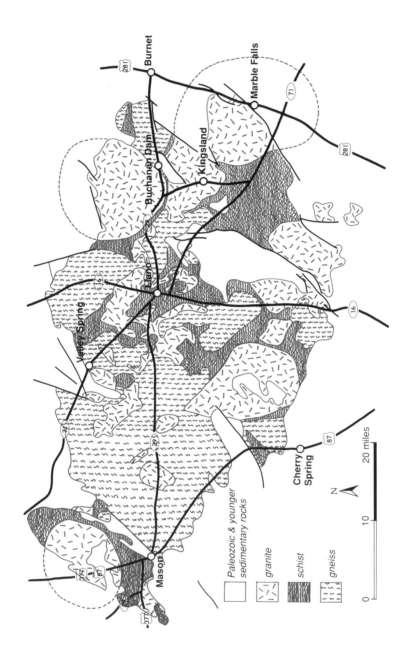

The Precambrian igneous and metamorphic rocks of the Llano uplift.

Paleozoic & younger
sedimentary rocks

granite

schist

gneiss

N

20 miles

Burnet

Marble Falls

Kingsland

Buchanan Dam

Llano

Valley Spring

Mason

Cherry Spring

U.S. 87
Fredericksburg—Mason
42 miles

The road between Fredericksburg and Mason crosses limestone terrain of the Edwards Plateau near Fredericksburg. Northward it crosses the western part of the Llano uplift, passing several excellent roadcut exposures of Precambrian gneiss and schist, and brown Cambrian sandstone.

North of Fredericksburg, U.S. Highway 87 crosses Hensell sand for three miles, to ascend on to the Edwards Plateau where creamy-colored rubble dots the plateau surface. Watch for several good roadcuts 10 to 15 miles out of Fredericksburg where the flat-bedded character of the limy beds, and their contained fossils can be observed. About three miles before reaching Cherry Spring, the road reaches the edge of the Edwards Plateau, where you get a panoramic view northward across the low terrain of the Llano basin, before the road winds downward on to Hensell sand. Gypsum was once mined west of the highway a short distance to the south, and is still mined from a gypsum-rich section of the Edwards limestone about four miles to the southeast.

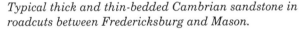

Typical thick and thin-bedded Cambrian sandstone in roadcuts between Fredericksburg and Mason.

In the low country between Cherry Spring and Mason, several excellent roadcuts are to be seen. At the Loyal Valley Road is a large roadcut in red-tan, Cambrian sandstone. Note the thick and thin horizontal beds and how the quartz grains stand out—if you look closely. This sand was deposited about 550 million years ago, well after life had developed a wide spectrum of forms on earth. Continuing along, skyline ridges on either side of the road are also of Cambrian sandstone.

Bouldery outcrops about 25 miles north of Fredericksburg are dark, platy, Packsaddle schist. Then, in another ten miles, just shy of Mason, schist and gneiss are exposed in quite spectacular roadcuts, where pegmatite dikes criss-cross the dark metamorphic rocks. The dikes were originally molten, watery liquids, squirted into fractures in the older metamorphic rocks. The dikes' parent liquid came from the huge, hot, molten bodies of granitic magma which rose through the metamorphic rocks nearly one billion years ago.

Just south of Mason, roadcuts and outcrops of Cambrian sandstone are seen again. In Mason, it is interesting to see how many historical buildings were constructed of this sturdy brown sandstone.

U.S. 87
Mason—Brady
28 miles

Between Mason and Brady, U.S. 87 traverses a section of Cambrian, Ordovician, and Pennsylvanian rocks which flank the older Precambrian rocks of the Llano uplift. The older Cambrian rocks are found near Mason, whereas the younger Pennsylvanian rocks occur near Brady. The sedimentary stack of these Paleozoic limestones and sandstones is thereby quite regular; arranged from south to north, the rocks are oldest to youngest, and bottom to top.

Small roadcuts of rather poorly exposed, brown Cambrian sandstone are in the northern part of Mason. Along the next four miles of highway north of town look for roadcuts where dark Packsaddle schist is cut by dikes, and for exposures of pink Town Mountain granite. Then, looking north, you see a skyline ridge of yellow-white limestone, which the road climbs and crosses. This is a prong of the high-standing plateau of Cretaceous Edwards limestone and Hensell sand, which directly overlie the Precambrian granite at this point. Excellent

About 6 miles north of Mason, US 87 cuts through a ridge of lower Cretaceous limestone and sandstone, which lay directly on Precambrian granite.

roadcuts expose thick and thin limy beds, though few fossils are to be found here. In the low area immediately north of the ridge, more pink granite appears along the roadside, its crystalline minerals sparkling in the sun.

Both north and south of the hamlet of Camp Air, the countryside has flattened. The exposureless area for two miles south of Camp Air is the lower part of the Hickory sandstone which is composed of well-rounded sand grains. Eight miles to the northeast it is quarried for hydraulic fracturing sand, used to enhance oil production. Two miles north of Camp Air the highway crosses the dark red or brownish-red upper part of the Hickory sandstone. The sand is red because it contains about 12% iron, and could be a potential source of iron in the future. In a quarry north of Camp Air, note the distinct sedimentary bedding of the brown sandstone.

Not many rock exposures are seen for the few miles south of the San Saba River crossing at Camp San Saba. The few outcrops are gray limestone of the uppermost Cambrian rocks, overlaid by Ellenburger group rocks of Ordovician age. And, from San Saba to Brady the highway crosses Ordovician strata, though not much more than rock rubble in the fields can be seen for geologic evidence. Brady itself lies on the very southern tip of a band of Pennsylvanian-age rocks, which form the bedrock over a wide area to the northeast (Northcentral section of this guide).

U.S. 90
San Antonio—Uvalde
70 miles

The eastern part of the road crosses the Medina Valley, the so called "Alsace of Texas," where rich soils yield abundant crops. The highway runs a few miles south of, and parallels, the main Balcones fault, which separates this downthrown area from the uplifted Texas Hill Country to the north.

U.S. 90 climbs over a north-south ridge just west of Castroville, where good roadside exposures of gravels and sandstones tell a story of Pleistocene stream deposition. Similar roadcuts are scattered along U.S. 90 for about ten miles. Look for white caliche deposits atop some of these cherty gravels. The chert is derived from chert nodules weathered from the Edwards limestone and transported here from the higher Edwards Plateau country to the north.

A local rock quarry can be seen on the ridge north of Hondo. At D'Hanis, fossiliferous clay pits, a brick factory, and red brick houses and buildings attest to the importance of the lowly substance of clay in our lives. West of D'Hanis the road climbs a gravel-capped terrace ridge out of the valley bottom.

In a roadcut three miles west of Sabinal look for dark gray-green volcanic rocks that are spheroidally weathered, and overlain by white caliche soil that has worked its way down between the volcanic

Dark volcanic breccia rocks weathered and invaded by white caliche soil west of Sabinal on US 90.

Quarry in basaltic volcanic rocks at Knippa, Texas on US 90.

fragments. Phenocrysts—large crystals of olivine mineral grains—have been weathered out leaving squarish holes where iron oxide (rust) has been emplaced, the iron oxide itself being a weathering product from high-iron content minerals in the volcanic rocks. Roadcuts of upper Cretaceous limestone and Pleistocene stream gravels continue to Knippa.

At Knippa, two prominent volcanic domes stand above the otherwise flat terrain north of the highway. Another of these volcanic domes is mined west of town in the Knippa quarry, by the White's Mines Division of the Vulcan Materials Company. U.S. 90 passes near the edge of the quarry, so you get a good opportunity to see these intriguing rocks from your car. The Knippa dome is part of a swarm of igneous bodies of late Cretaceous age (average age about 80 million years) that occur in a belt which trends eastward from Del Rio nearly to Waco. The belt roughly parallels the edge of the Cretaceous Edwards Plateau and the Balcones fault zone. Most of these bodies were submarine volcanic centers where eruptions created explosion craters on the Cretaceous seafloor. The lava came up along late Cretaceous faults, which follow the familiar Balcones fault trend. Rocks in the Knippa quarry are dark-colored nephelinite, a sodium-rich rock similar to basalt. At Knippa, actual fragments of mantle rock are commonly found, so these extrusions came from deep within the earth, from the mantle below the earth's crust. There must have been

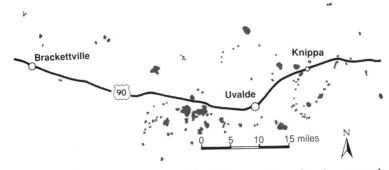

Volcanic eruptions blasted skyward in Cretaceous time, leaving a number of lava bodies in the Uvalde area.

quite a Cretaceous show when these volcanoes erupted in the sea, spewing steam, hot rocks, and red lava skyward.

From Knippa to Uvalde, the road continues to traverse relatively flat terrain where occasional upper Cretaceous limestone and claystone exposures peek from the extensive cover of the surrounding river deposits. At the town of Uvalde, a prong of upper Cretaceous rocks extends southward toward town from the north, allowing Edwards Limestone spring water to emerge near Uvalde, where the fresh water is used to irrigate the adjacent agricultural fields. Uvalde is the former home and now the gravesite of John Nance Garner, who served as Vice President of the United States under President Franklin Delano Roosevelt.

U.S. 90
Uvalde—Del Rio
72 miles

The road traverses mostly upper Cretaceous limestone, with noticeable intrusive volcanic hills about halfway between Uvalde and Bracketville.

At the crossings of the Nueces River and Sycamore Creek, excellent views of sandy, meandering river point bars can be seen in the river bends.

Note another set of quarries where volcanic rocks are mined about 12 miles west of Uvalde on the hills south of the road. In an area south of Cline and Blewett, though unseen from US 90, are old asphalt mines from the turn of the century. These were the largest natural asphalt mines in Texas, and perhaps in the United States. The asphalt is from natural oil seeps which impregnated a coarse, shelly limestone. The asphaltic limestone was mined, crushed, and laid down "as is" on roads. Highway F.M. 1022 to Blewett is composed of this "natural pavement."

A noticeable change in vegetation occurs along this road stretch. Temperate trees, shrubs, and grasses give way to desert topography and plants about 20 miles west of Uvalde. Desert terrain is more and more prevalent west from here.

The flat-lying upper Cretaceous limestones, uninterrupted by canyons or stream cuts, account for the mainly flat terrain between Uvalde and Del Rio. White limestone rubble in fields adjacent to the highway is all you see in many places. Just east of Del Rio, note the white caliche beds in the roadside exposures near the entrance to Laughlin Air Force Base. In Del Rio, a major Edwards aquifer springs, called San Felipe Springs, bubble up in the golf course in town. The springs feed the San Felipe River, which empties into the Rio Grande.

U.S. 90
Del Rio—Langtry—Sanderson
120 Miles

This stretch of West Texas desert country forms the western margin of the Edwards Plateau. The geologic story, therefore, between Del Rio and Sanderson, is expectedly one of Cretaceous limestone terrain, but a decidedly interesting one, especially topographically. For here, thick sequences of limy rocks are deeply dissected by the Devils' River and the Pecos River, which carve their way southward to their inevitable junctures with the Rio Grande. Magnificent cliffs stand tall above these rivers, and scenic highway crossings are found at Amistad Reservoir (Devils' River) and at the spectacular high bridge over the Pecos River.

The U.S. 90 roadbed around Del Rio rides on the surface of a thick, gray, lower Cretaceous limestone which is beautifully exposed to the west in the Pecos River Canyon. This limestone is also the same one that forms the upper part of the giant cliffs in Santa Elena Canyon on the west side of Big Bend Park. It is indeed a widespread unit.

Geologic map of Del Rio area.

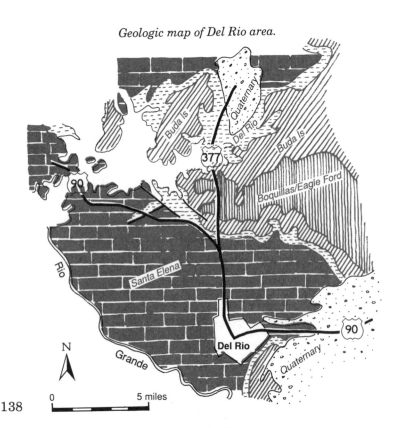

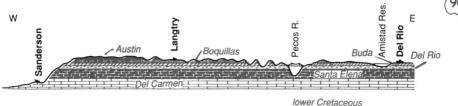

Cretaceous section along US 90 between Del Rio and Sanderson.

Two miles north of Del Rio, U.S. Highway 377 splits off to the right, heading north to the Rough Canyon Recreation area of the Amistad Reservoir. As a short five mile side trip, U.S. 377 between this junction and the lake crosses a ridge where several roadcuts slice through the upper Cretaceous Eagle Ford, Buda, and Del Rio formations, as shown on the geologic map. The Eagle Ford (called "Boquillas" farther west of here) is mainly thin-bedded ("flaggy") limestone, alternating with shale and siltstone. In a quarry two miles north of the U.S. 90—U.S. 377 junction, the Eagle Ford is mined for crushed stone and aggregate.

The sequence of rocks in the Del Rio — Sanderson area.

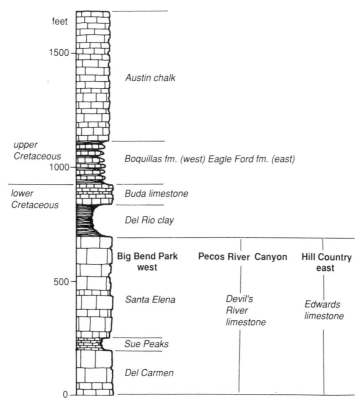

Hard, nodular, brownish Buda limestone, loaded with clam fossils, is seen in cuts just south of the Lake Amistad Bridge. The Del Rio clay is not easy to spot, but it is a yellowish gray, limy claystone and siltstone just below the Buda limestone.

Back on U.S. 90 west of Del Rio, the road crosses the bridge over Lake Amistad, which dams the Devils' River and Rio Grande. You can take Spur 349 off U.S. 90, twelve miles west of Del Rio for a short, three mile side trip to see the dam and the broad canyon of the Rio Grande.

West of the U.S. 90 bridge over the lake, natural outcrops of weathered gray, lower Cretaceous limestone form scenic cliffs and bluffs.

About 35 miles west of Del Rio, west of Comstock and Cow Creek, several good roadcuts show hard, light-gray, thin-bedded Buda limestone, full of clams and burrows, overlying brown, clayey beds of the Del Rio clay, which contains abundant fossil "hash." Look for Ramshorn clams in the Del Rio here.

Thin Eagle Ford (Boquillas) beds rest on white Buda limestone beds.

140

Lower Cretaceous limestone, west side of Amistad Lake Bridge on US 90.

White Buda limestone over brown Del Rio clay a few miles west of Comstock on US 90.

In a few miles, the highway approaches the high bridge over the Pecos River. Boquillas, Buda, and Del Rio rocks are beautifully exposed in cliffy roadcuts on the eastern approach to the bridge. Watch for a blue sign that says "Picnic Area—Scenic Overlook—Historical Marker," which marks the entrance to a gorgeous scenic overlook on the south side of U.S. 90. From the overlook, the deep, cliff-walled incision of the Pecos River canyon spreads north and southwest, while the Pecos-Rio Grande confluence can be seen on the hazy horizon to the southwest. The lower Cretaceous Devils' River limestone comprises most of the cliff walls down to river level. This limestone is equivalent to the Sue Peaks—Santa Elena—Del Carmen limestone that form the famous cliffs in Santa Elena Canyon at Big Bend National Park. The upper Cretaceous beds on top of the Devils' River limestone form the uppermost cliff edge, which we have seen in the highway roadcuts on the bridge approach. One can only marvel here at the power of erosion and the persistent downcutting by the Pecos River over the last few million years as it chipped away at these hard limestones to create this beautiful canyon. Others before us have also enjoyed this canyon, for abundant living sites, rock art, and lithic fragments of the ancient Pecos River culture litter the canyon walls and floor.

Crinkled, flaggy Boquillas beds over white Buda limestone, 6 miles west of Comstock on US 90.

Approaching the colorful, historical town of Langtry, where Judge Roy Bean once held legal sway in West Texas, watch for crinkly, gray to red, nodular, thin-bedded limestone in roadcuts. These beds are upper Cretaceous Boquillas formation, quite spectacular in their odd nodularity. And, of course, you will want to leave geologic pursuits for a while to soak up historical Texana at Langtry.

Highway U.S. 90 climbs westward from Langtry and in a few miles tops out on a plateau where excellent roadcuts reveal shockingly white, blocky mudstone, and chalk beds of the upper Cretaceous

Ramshorn snails are common in Buda/Del Rio rocks. The quarter (right) and dime (below) are used as a scale.

Austin chalk. It is surprising how recognizable the Austin is; because it looks just the same here as it does around Austin and as far away as Dallas! Clams and their thick calcite shells are common in the Austin chalk.

The highway rolls along on the Austin chalk for about 30 miles, the plateau-like countryside being punctuated by Lozier Canyon, about 18 miles west of Langtry. There the Austin is breached and underlying Boquillas and Buda rocks form the lower walls in a natural river cut south of the road. The Austin-supported plateau east of Dryden gives way westward to valley and mesa country, cut into Boquillas and Buda limestones by satellite creeks and draws of the Rio Grande.

Eight miles west of Dryden, U.S. 90 begins its descent into Sanderson Canyon, a delightful drainage, cut deeply into lower Cretaceous limestone by Sanderson Creek, which is a tributary of the Rio Grande lying twelve miles to the south. Rocks north of the road are mainly cherty, fossiliferous Edwards limestone (lower Cretaceous), whereas south of the highway equivalent limestones are called Santa Elena. Such terminology is used in West Texas and Big Bend Park for this significant assemblage of cliff-forming limestones. The canyon deepens and darkens as it narrows toward the town of Sanderson, where we leave this section and continue westward on U.S. 90 in the West Texas section.

The road between Sanderson and Del Rio has traversed many of the upper and lower Cretaceous limestone units which make up the bulk of the Edwards Plateau. The roadcuts and natural outcrops are many, and give ample opportunity to see these rocks first-hand and up close. The vast expanse of Cretaceous limestones is breached westward around Marathon, Texas and in the Big Bend country, and we won't see these rocks again until we visit the Sierra Del Carmen Mountains on the east side of Big Bend National Park and Santa Elena Canyon on the west side of Big Bend Park. There the sequence is seen in its entirety in sheer canyon walls.

West Texas desert mesa of lower Cretaceous Santa Elena limestone, 2 miles east of Sanderson on US 90.

White even-bedded limestone of the Austin chalk formation, 6 miles west of Langtry on US 90.

Pecos High Bridge over Pecos Canyon looking north. Thick, lower Cretaceous limestone (Devils' River limestone) forms steep lower cliffs.

U.S. 183
Lampasas—Austin
63 miles

Highway 183 between Lampasas and Austin crosses the southern end of the topographic region known as the Lampasas Cut Plains, which forms the eastern Hill Country edge of the uplifted Edwards Plateau.

In Lampasas, note the number of historical buildings built of locally-derived limestone blocks. It is often easy to get an idea about the geology of an area by examining the stonework in pioneer buildings!

Lampasas lies in the Sulfur Creek drainage where erosion has cut completely through the lower Cretaceous Edwards and Glen Rose formations exposing Pennsylvanian Marble Falls limestone and Ordovician Honeycut formation in the river bottom west of town. However, these older rocks are not seen along U.S. 183.

South of town, U.S. 183 climbs through yellowish Cretaceous limestone beds of the Glen Rose limestone. Look for exposed rocks in roadcuts and natural outcrops on either side of drainages, such as Mesquite Creek and Rocky Creek. About 20 miles southeast of Lampasas, the road climbs noticeably to the top of a plateau where Edwards limestone forms the flat surface. Lower, older Glen Rose is seen again, however, east of the bridge over the North Fork of the San Gabriel River. The road again climbs onto the Edwards surface around Seward Junction, then heads downward to cross the South Fork of the San Gabriel River. Square patterns of joints can be seen in the flat Glen Rose beds in the river bottom.

Around the town of Leander, the highway is again on the flat Edwards surface on which it continues on into Austin. Suburb construction becomes predominant toward Austin, so not many rocks are to be seen. The road passes over the Balcones fault system on the northwest suburban outskirts of Austin and crosses the Mount Bonnell fault, the western main fault of the Balcones fault system, about where the Capital of Texas Highway (Spur 360) joins U.S. 183. Please refer to the section on the geology of Austin to complete the story here.

U.S. 281
Marble Falls—Johnson City
23 miles

Along U.S. 281 between Marble Falls and Johnson City, older Cambrian and Ordovician rocks lie directly beneath younger Cretaceous rocks. The older rocks are seen where stream erosion has cut through the Cretaceous cover.

At the south end of the U.S. 281 bridge over the Colorado River, look north to see the cliff of inclined Marble Falls limestone of Pennsylvanian age in the river bank below town. These dark gray limestones continue for a mile south of the river and form small roadcuts along U.S. 281.

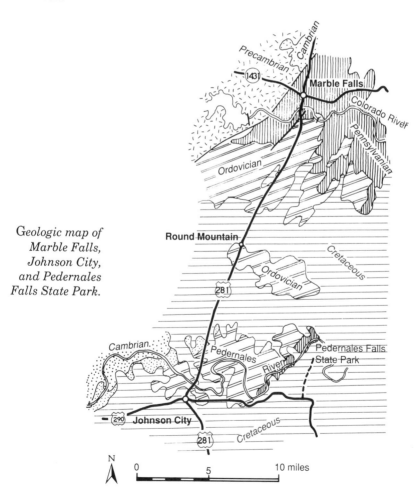

Geologic map of Marble Falls, Johnson City, and Pedernales Falls State Park.

For seven miles farther south, Ordovician Ellenburger group limestones (Ordovician) appear in small cuts beside the road. South of the Shovel Mountain community, the road climbs a low drainage divide where soft yellowish Cretaceous limestone and sandstone beds are found. These Cretaceous rocks are breached around the hamlet of Round Mountain; coarse-grained Cambrian dolomite and siltstone are the roadside rocks in the lower elevations for a few miles south of Round Mountain.

The highway climbs another drainage divide and passes through about six miles of Cretaceous Hensell sand and Glen Rose limestone beds. Four miles north of Johnson City the road once again traverses onto Ordovician rock terrain, then passes a patch of Cretaceous Hensell sand north of the Pedernales River (locally pronounced "Perd-n-Alice"). A mile north of Johnson City, you cross the Pedernales River where hard, gray Ordovician dolomites are well exposed by river erosion. Note how the Paleozoic rocks along this road are tipped at an angle to the southeast, reflecting their position on the flank of the Llano uplift. Excellent exposures of tilted Paleozoic rocks can be seen in Pedernales Falls State Park, five miles east of Johnson City.

Sauropod tracks in the Blanco River west of Blanco, Texas.

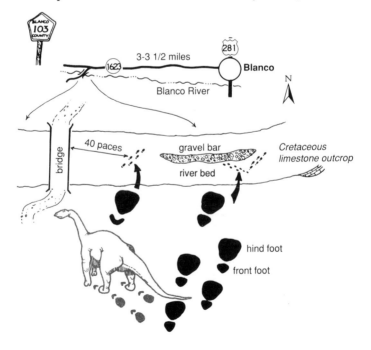

Erosion rills on lower slopes of roadcut in lower Cretaceous Glen Rose formation on US 281 north of Blanco, Texas.

U.S. 281
Johnson City—Blanco
14 miles

The highway between Johnson City and Blanco passes low hills where yellow-tan limestone ledges peek out every now and then from the brush country. All the rocks are lower Cretaceous Glen Rose limestone, whose thin-bedded, alternating marl and limestone depositional style creates stair-stepped topography, which is displayed all along this road.

One hill a few miles north of Blanco shows the stair-step character very well. Also look along the roadway for erosion rills in the soft, yellowish marly layers of the Glen Rose. Fossil "hash" of clams and snails is quite common in these shallow marine beds.

A delightful short side trip can be made from the town of Blanco to view tracks of sauropod dinosaurs. Go west from Blanco for about 3 miles on F.M. 1623, which follows the Blanco River. Look carefully for the Blanco County Road 103 sign, and turn left. The road immediately crosses the Blanco River where sauropod tracks are preserved in the limestone beds in the river bottom. The map shows where the tracks are found. These bathtub-sized tracks would be easy to overlook as tracks, except for their regular spacing, biological symmetry, raised edges (where mud punched up around the foot), and size, which matches the foot and step length of sauropods as determined from skeletons and other trackways. Moreover, tracks in the Glen Rose are common over a wide area in Texas, so these are certainly not unique.

149

U.S. 290
Austin—Johnson City—
Fredericksburg—I-10
118 miles

Highway U.S. 290 west of Austin is an east-west traverse of the Hill Country and Edwards Plateau. The rocks seen along the way are, with rare exception, Cretaceous in age, including the blocky Edwards limestone, the thin-bedded Glen Rose limestone, and the Hensell sand. This uplifted pile of essentially flat-lying rocks rests on a planed-off surface of much older Paleozoic rocks.

Beginning in Austin, U.S. 290 heads southwest from its in-town junction with I-35. The road between I-35 and the small town of Oak Hill, located a few miles southwest of Austin, traverses down-faulted Edwards limestone in the Balcones fault zone. The Mount Bonnell fault is the main bounding fault of the Balcones fault zone, and runs north-south through Oak Hill. The rocks west of Oak Hill are uplifted, Cretaceous Glen Rose limestone, which is famous across Texas for its abundance of Dinosaur tracks.

In the road curve west of Oak Hill is a large bluff of Glen Rose rocks. The alternating thin beds of hard limestone and soft marl in the Glen Rose create its distinctive 'stair-step' topography seen throughout the Hill Country.

From Oak Hill to Johnson City, U.S. 290 rides entirely on Cretaceous Glen Rose rocks. Watch for a quarry wall exposure on the north side of the road in the town of Dripping Springs.

Geologic cross section along US 290

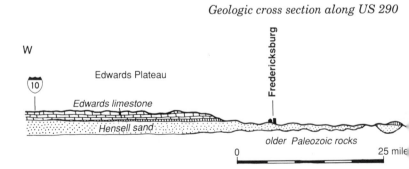

The hilly, eroded topography of the Hill Country is due to deep erosion by streams which have been attacking the edge of the Edwards Plateau since its uplift in Miocene time, ten million years ago.

Johnson City is located on the edge of the Cretaceous Hill Country terrain; a short drive north of town on U.S. 281 takes you into older rocks of Paleozoic (Ordovician) age.

Highway U.S. 290 follows the Pedernales River (locally pronounced "Perd-n-Alice"), made famous by President Lyndon B. Johnson, whose ranch, now open to the public, is located about fifteen miles west of Johnson City. Though the majority of rocks around the LBJ ranch are lower Cretaceous Hensell sand, older rocks of Cambrian age crop out beneath the Cretaceous cover in the Pedernales River bottom.

The low-lying countryside between the LBJ ranch and Fredericksburg is underlaid by Cretaceous Hensell sand. The Pedernales River here has, therefore, eroded completely through the Edwards limestone, Glen Rose limestone, and into the Hensell sand.

The town of Fredericksburg is a delightful, historical Texas town where 19th Century German immigrants built solid houses and shops of locally-derived limestone blocks.

West of Fredericksburg, U.S. 290 travels along for a few miles on low river-cut terrain. It then climbs the edge of the Edwards Plateau to emerge on the Edwards limestone, a characteristically blocky, thick-bedded, light gray limestone and dolomite, along with gypsum in some areas. Fairly flat Edwards countryside continues for the remainder of the way to the U.S. 290 junction with Interstate 10.

between Austin, Fredericksburg, and I-10.

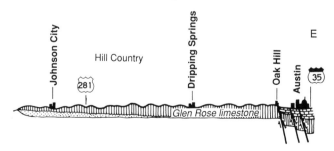

Texas 16
Kerrville—Fredericksburg
24 miles

Texas 16 heads northeast from Kerrville to Fredericksburg, crossing Interstate 10 north of town. On the flanks of the Guadalupe River
drainage near Kerrville, yellowish, lower Cretaceous Glen Rose limestone beds form the lower slopes of cliffs and roadcuts. Away from the
river, Texas 16 traverses about five miles of a flat plateau surface
formed on the Edwards limestone. Farther north, the highway descends through the Glen Rose limestone as it follows Wolf Creek on its
way across the Pedernales River and to Fredericksburg. Between the
crossing of the Pedernales River and Fredericksburg, Texas 16 is on
the Hensell sand, deposited at the same time as the Glen Rose
limestone to the east. About three miles from Fredericksburg watch
for a roadcut across from the entrance to Lady Bird Johnson Munici-
pal Park and Campground. Here, beautiful cross-bedding in conglom-
eratic sandstone, deposited in an earlier stage of the Pedernales
River, is characteristic of stream channel deposition.

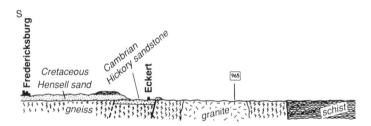

Texas 16 cross section between Fredericksburg,
granite, and schist predominate in the

Texas 16
Fredericksburg—Llano
39 miles

North of Fredericksburg, Texas 16 traverses Cretaceous rocks, crossing about eight miles of poorly exposed, flat-lying, thin beds of tan-colored Hensell sand seen in hills on either side of the road. A nice view north into the Valley of Palo Alto Creek comes up at about 2 miles north of town. At eight miles is a high-standing prong of the Edwards Plateau, where the road crosses about a mile of Edwards limestone, then descends toward the Llano basin proper, dropping through a short segment of Hensell sand again. Watch carefully for brown sandstone roadcuts for about a mile on either side of the small hamlet of Echert. This is Cambrian Hickory sandstone preserved in a faulted, down-dropped block on the edge of the Llano uplift.

Look west to see the topographic edge of the Edwards Plateau, where the Cretaceous sandstone and limestone form a distinct cliff.

North of Echert, hills scattered with pink, blocky boulders of gneiss and granite are readily seen from the highway, signalling entry into

Llano, and San Saba. Precambrian gneiss,
center of the section.

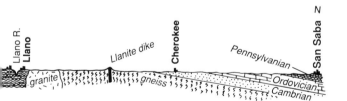

the central Precambrian portion of the Llano uplift. Looking east from the junction of Texas 16 and F.M. 965 (the side road to Enchanted Rock State Park) the prominent skyline hills are a block of Paleozoic sedimentary rocks, which occupy a graben (down-dropped fault block) that was dropped into the Precambrian rocks at the southeast corner of the Llano uplift. The Paleozoic sedimentary rocks were actually more resistant to weathering than most of the surrounding Precambrian crystalline rocks, so they stand elevated here as a ridge line.

North of the F.M. 965 junction to Llano, Texas 16 is in Valley Spring gneiss and Packsaddle schist terrain. Watch for low roadcuts where the light-colored gneiss looks like contorted, banded granite, whereas the schist is black and has platy minerals that sparkle in the sunlight. From the bridge over the Llano River in the town of Llano, views of sand bars are seen during low water periods, and Valley Spring gneiss and Packsaddle schist boulders and outcrops are nicely exposed in the river bottom.

Texas 16
Llano—San Saba
33 miles

This road segment offers the opportunity to view the igneous and metamorphic rocks, some quite unusual, of the Llano core, as well as the beds of hard, Paleozoic sandstone and limestone which flank the uplift's crystalline center.

Low, relatively featureless topography is encountered for eight miles north of Llano, though pink granite and gneiss boulders are seen here and there. Nine miles north of town, Baby Head Hill is part of an east-west ridge which is held up by a hard igneous dike that stands

Sketch of a piece of Llanite — an unusual granite. Feldspar minerals are red/pink, quartz grains are blue. Pencil color the round white areas blue, and the sketch will resemble the actual rock.

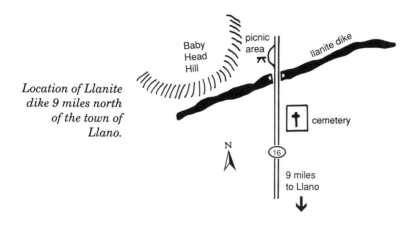

Location of Llanite dike 9 miles north of the town of Llano.

Baby Head Hill

picnic area

llanite dike

cemetery

N

16

9 miles to Llano

above the surrounding Valley Spring gneiss. The dike can be seen in roadcuts on either side of the highway where the road cuts the ridge. The dike is composed of a striking granite which contains red-pink feldspar crystals and blue quartz grains that float in a fine-grained, almost black groundmass. The rock is unusually hard, and was quarried at one time for building stone. The quartz is blue because of chromium impurities. Unique to this area, the rock is appropriately called Llanite. You really must get out of your car and examine this rock up close in the roadcut.

Farther north, the road passes from Precambrian igneous terrain into the fringe belt of Paleozoic sandstone and limestone. The Paleozoic rocks dip gently to the north and, from south to north, are stacked in proper sequence, oldest to youngest, Cambrian to Pennsylvanian. These sedimentary rocks were deposited on the bevelled surface of Precambrian rocks. Watch for a roadcut on the right about four miles north of the Llanite dike, where the first Cambrian sandstone (Hickory sandstone) is exposed. Here, the gray-tan, pebbly, and very hard sandstone is shot through with thin calcite veins. This sandstone was laid down directly on the flat, eroded, Precambrian surface.

The landscape around the town of Cherokee is open and quite flat, but look for a quarry on the west side of the road a few miles north of town, where hard, glinty blocks of dark gray Cambrian–Ordovician limestone here are about 500 million years old.

A few miles farther, in the stream valley of Buffalo Creek, another very hard limestone crops out, the Ordovician Ellenburger group. Note it too is very dense, and has many chert nodules in it. The nodules are thought to be redistributed silica, derived from original layers of siliceous marine organisms, such as sponges and diatoms.

Abundant limestone boulders, weathered to a battleship gray are lying out in the fields, derived from the underlying beds of Ordovician-age rocks. At Simpson Creek crossing, just south of the town of San Saba, thin limestone beds in the roadcuts are Pennsylvanian, a little over 300 million years old. This is Marble Falls limestone, named for ledges exposed east of here at the town of Marble Falls.

This road segment ends at the town of San Saba, the pecan capital of Texas.

Texas 29
Llano—Mason
34 miles

This east-west drive traverses the very heart of the Llano uplift, rolling along almost entirely on the billion year old granite, gneiss, and schist complex of rocks that characterize the Llano country. The topography is by no means spectacular, but here and there along the way, low rounded knolls of pink granite, and banded slabs of gneiss form the platform for a clump of trees, almost emulating a small scene from a Japanese garden. To the east, knobs and knolls coalesce in skyline ridges, reminding the observer that erosion may have levelled once-tall mountains, but it left a bit of topography on the remnant surface.

Knolls of pink granite dot the countryside between Llano and Mason.

Contorted Packsaddle schist in roadcut east of Mason on Texas 29. This rock is over 1 billion years old.

Just east of Mason, watch for a sizeable roadcut north of the highway, where grayish bands of gneiss and schist (Packsaddle schist) are lanced by lighter-colored granitic dikes.

Historical buildings in Mason, many of which are well-preserved, are delightfully constructed of locally-quarried brown, Cambrian-age Hickory sandstone. This sandstone is preserved on the edge of and in the Llano uplift in elongate, down-dropped blocks between linear faults.

Historical buildings in Mason are commonly constructed of locally quarried Cambrian sandstone.

Texas 29
Llano—Buchanan Dam—Burnet
30 miles

The road between Llano and Buchanan Dam follows a large bend in the Llano River for part of the route, traversing Precambrian crystalline rocks the entire distance. Near Llano the road is mainly on gneiss, then on Town Mountain granite from the river bend to Buchanan Dam. Gneiss is again the main bedrock between Buchanan Dam and Burnet, but Cretaceous sandy sediments and limestone are encountered near Burnet.

A few miles east of Llano, granite and gneiss outcrops near the highway appear as low platforms, protruding above the grasses in surrounding fields. Note especially how jointed or fractured the rocks

Geologic map of the east side of Llano uplift in the Marble Falls area.

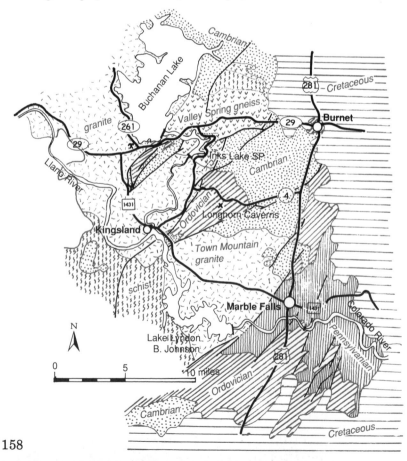

Fractured granite east of Llano. The granite also peels off in sheets by exfoliation.

are here. This is also a chance to see exfoliation sheets at close range, if you did not or do not get the chance to see them while hiking around Enchanted Rock State Park. Very coarse, crystalline dikes of pegmatite snake through these rocks as well.

Watch for well exposed outcrops and boulders of Packsaddle schist south of the highway in the Llano River bed where the road hugs the river bank.

Coarse, crystalline granitic dike (pegmatite) cuts across schist. The quarter is for scale.

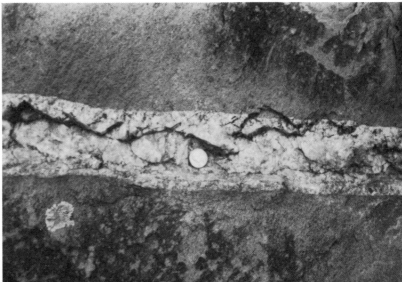

Blocks of granite in a quarry along FM 261 west of Lake Buchanan. Look for zoned feldspar crystals.

As the highway nears Buchanan Dam, watch for the exit north to F.M. 261 on the west side of Lake Buchanan. About 1/4 miles north of Texas 29 on F.M. 261 is a roadside quarry where large, zoned feldspar crystals stand out of the pink granite. The feldspar zoning represents a change in chemistry of the cooling magma, such that the composition and hence color, of the feldspar crystals changed as they grew in the hot, liquid melt.

From Buchanan Dam east to Burnet, most of the road is on Valley Spring gneiss. The edge of the Llano uplift is found near Burnet, where Cretaceous, sandy sediments, overlain by limestone, rests directly on the erosional surface, on Paleozoic and Precambrian rocks.

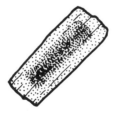

Zoned feldspar crystal.

Texas 46
New Braunfels—Boerne
43 miles

Classic terrain of the Edwards Plateau country is crossed in this road segment. Limestone crops out everywhere. Piles of limestone rock at roadsides, and roadcuts through vertical faces of limestone are common. Sparse vegetation and nearly mesa-like hills are typical. The entire road segment is on lowermost Cretaceous rocks. Limestone, sandy marlstones, shales, and calcareous sandstone typify the rocks seen along the way. Edwards limestone and Glen Rose limestone are noted around the state for their fossils of marine snails, clams and dinosaur footprints.

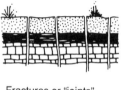

Fractures or "joints"
(no motion)

Faults
(motion)

Just west of the Guadalupe River, Texas 46 cuts through two very nice roadcut exposures. Plenty of space to park here and easy access to outcrops make this a worthwhile stop to look at the rocks. See especially the rubble of limestone on the north side cut. This is rock which once collapsed into underlying limestone caverns. Such terrain is called "karst," where sinkholes, caverns, and dead-end surface drainages which end in sink holes are characteristic. The southside cut is highly fractured or "jointed" - note the high angle, nearly vertical "joints." Joints are fractures which show no relative motion across the fracture. Faults are fractures where one side moves relative to the other side. Molds of fossil clams and snails are seen in the hard, dense gray limestones on the southside walls.

Natural Bridge at Entryway to Natural Bridge Caverns. Note collapse blocks below bridge.

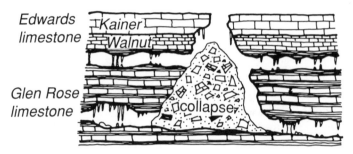

Edwards limestone

Kainer

Walnut

Glen Rose limestone

collapse

Geologic cross section of Natural Bridge Caverns.

Watch for signs to Natural Bridge Caverns a few miles west of New Braunfels. This is a beautiful cave with some of the best dripstone features of any cave in the state. (Refer to section on Texas Caverns.)

Farm Road 965
Fredericksburg—Enchanted Rock
State Park—Texas 16
26 miles

This is a road of visual and geologic surprises! From Fredericksburg, the highway heads north across clayey, silty, sandy Cretaceous sediments to the limestone terrain of the Edwards Plateau. Lower Cretaceous limestones, deposited in the shallow seas that once spread across Texas 100 million years ago, form hillside ledges all around. A few miles north of town a vista to the north comes into view; to the left, west, is a ridge of flat-bedded limestone, but to the right, east, is a dark red, bouldery hill looking distinctly out of place. This steep-sided hill is Bear Mountain, which is composed entirely of pink granite, which is mined here for architectural stone.

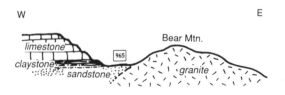

W

E

Bear Mtn.

limestone

claystone

sandstone

965

granite

Bear Mountain north of Fredericksburg.

162

The granite knob appears to be at about the same level as the limestone across the road. Did the granite penetrate the limestone? Or was Bear Mountain a huge knob on the Cretaceous seafloor against which the limestone was piled? To answer this question, we need to know the ages of the two rocks, limestone versus granite. As it turns out the granite crystallized a billion years ago, while the limestone is "only" about 100 million years old. So the granite being much older, could not logically have penetrated the limestone. That means the granite knob must have been a hill on the Cretaceous seafloor, and the limestone beds lapped onto it, eventually covering it. Erosion has removed the relatively softer limestone from around the granite, which now stands high as it continues to resist erosion. Stop at the picnic area north of Bear Mountain to get a close look at blocks of this beautiful, fine-grained granite.

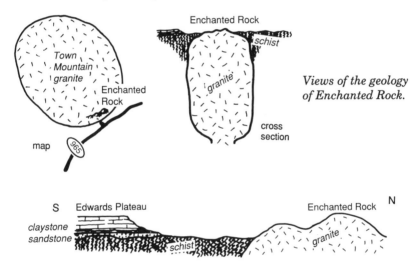

Views of the geology of Enchanted Rock.

The road continues north in hill-and-valley limestone terrain for another ten miles, only to turn an abrupt corner, where a gorgeous panoramic view of Enchanted Rock and the surrounding countryside is laid out from skyline to skyline. With an unimpeded view, the road twists downward across the topographic edge of the Edwards Plateau, toward the Sandy Creek valley below. It is clear here how the Cretaceous sediments were laid down directly on the surface of the Precambrian granites. But, it is also clear that quite a bit of topography existed on that surface, considering the large granite domes seen ahead. As the road crosses over Sandy Creek, look for black boulders in fields and along fences. These rocks are black schists which are the metamorphosed products of the shaly part of the original Packsaddle sediment wedge. These metamorphic rocks are full of platy minerals which formed when the shale was intensely heated far below the surface.

The road continues on Packsaddle schist for a few more miles, paralleling the granite domes, which loom ever-larger as you approach the entrance to Enchanted Rock State Park. Looking eastward at the junction with Texas 16 a few miles ahead, the prominent skyline ridge ahead is a fault-block of Paleozoic rocks that has been downdropped into the Precambrian in the southeastern part of the Llano uplift.

ENCHANTED ROCK STATE PARK

Enchanted Rock State Park is a marvelous piece of Texas landscape, where geologic features are clearly laid out in raw-rock profusion, devoid of soil or vegetation—or concrete cover. The view from the summit of Enchanted Rock, elevation 1,825 feet and 445 feet above Sandy Creek, is a gorgeous 360° panorama. Most obvious at first view of the park is the dome shape of the rocks. However large as they may seem, Enchanted Rock's granitic domes are but a small side piece of a huge, round globe of granite which rose through the Packsaddle schist like a giant, hot balloon about a billion years ago. The circular shape of the granite body, a batholith, can be seen on the maps, but is too big to make out from road level. The overall geologic history of the Llano country, which includes Enchanted Rock, is discussed in the Llano uplift section.

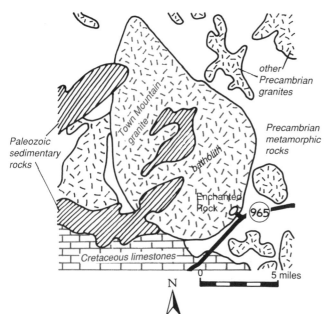

Geologic map of Enchanted Rock batholith. Enchanted Rock itself is only a small part of the entire batholith.

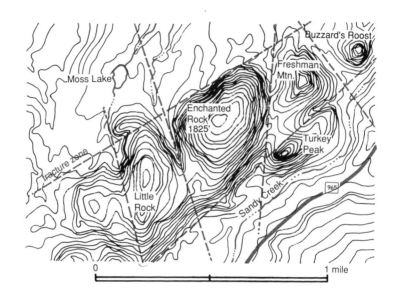

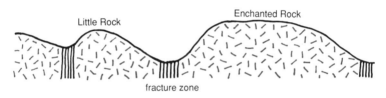

Fracture zones in granite control the topography at Enchanted Rocks.

sheets, creating blocks and slabs, which slide downslope. Exfoliation sheets and blocks are beautifully displayed on the north flank of Little Rock, and can be seen from the flanks of Enchanted Rock.

As you climb around, look for odd-shaped rock pedestals, created by the differential disintegration of the wetted, lower portion of boulders, where chemical weathering attacks the rock faster than the dry boulder top.

Note also the round weathering pits, or pools, where standing water causes chemical weathering to disintegrate the granite. These pits, filled with water after rains, are important water sources for animals in many desert areas of the world.

The surface of Enchanted Rock is crossed by long linear bumps and rills. These are caused by dikes which filled cooling cracks in the early granite batholith. As it crystallized, the granite shrank, cracks developed, and hot liquid from the last phase of the magmatic mush rushed

Cross fractures along Echo Canyon Trail in Enchanted Rock State Park.

A good example of exfoliation is displayed at Enchanted Rock.

167

Pedestal rocks at Enchanted Rock State Park.

Ponds and rills on top of Enchanted Rock.

in to fill the cracks. If the filling material is finer-grained than the surrounding granite, it is slightly harder, so it stands up to erosion forming a little ridge. But, if the granitic dike material is coarser grained, it weathers more easily, creating a little depression in the surface rock. These dikes are particularly noticeable on the face of Enchanted Rock.

Many other fine points of erosion, weathering, and granite rocks are to be seen by the careful observer. For those interested in more detailed information to help the search, purchase the well-written and illustrated pamphlet by James F. Peterson, entitled *Enchanted Rock State Natural Area – A guide to the Landforms*, published in 1988 by Terra Cognita Press, San Marcos, Texas. The book is available for a few dollars at the State Park entry building. And, while you are there, spend some time viewing the displays of Enchanted Rock geology. They are extremely well presented and informative.

Differential erosion on a pegmatite dike cutting granite at Enchanted Rock State Park.

Park 4, Farm Road 2342, Farm Road 1431
Buchanan Dam—Inks Lake State Park—
Longhorn Cavern State Park—Marble Falls
46 miles

This short drive is one of the most fascinating in the state for the variety of rocks to be seen and the excellent outcrops. Igneous, metamorphic, and sedimentary rocks are all exposed along this stretch, and bona-fide faults can be seen on Backbone Ridge. The wonderful, historical Longhorn Cavern is on the way, and granite quarries that date back to the 1880's, where the building stones for the Texas State Capitol come from, are also found along this road.

Turn south onto Park Road 4 from Texas 29 east of Buchanan Dam, which takes you to Inks Lake State Park. A couple of miles after making the turn, stop at the Devil's Waterhole overlook where you can

Map and cross-section of fault wedge along Backbone Ridge on the road to Longhorn Cavern.

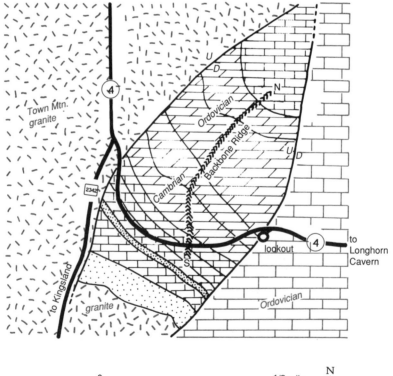

*Geologic cross
section of Backbone
Ridge.*

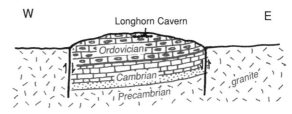

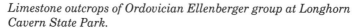

look down into the rough terrain carved into the Valley Spring gneiss by Spring Creek. The Valley Spring gneiss, more than a billion years old, formed originally from nearly remelted rhyolitic volcanic rocks, ash-flow tuffs, and tuffaceous sediments. They were deeply buried by sediments, followed by metamorphism during an episode of Precambrian mountain building. The knobby, irregular, rough terrain seen here is quite typical of the way the Valley Spring gneiss weathers. The scenic drive through Inks Lake Park weaves past many colorful, pink gneiss boulders, and natural outcrops.

At the south end of the park, the road crosses onto granite rocks in Hoover Valley; Backbone Ridge forms the skyline hills to the east. Watch for a sharp left turn of Park Road 4, which heads east toward Longhorn Cavern as it climbs the flank of Backbone Ridge. Immediately

*Limestone outcrops of Ordovician Ellenberger group at Longhorn
Cavern State Park.*

View west over Lake LBJ and Hoover Valley. Llano country in the distance, from Hoover Point on Backbone Ridge, FM1431 near Kingsland.

after making the turn, notice that the geology changes abruptly from pink granite to gray, hard, bedded limestone. You have just crossed a fault that marks the edge of a large wedge-shaped block of early Paleozoic sedimentary rocks preserved in the surrounding igneous terrain. The limestones in the exposures along the ascending roadway are almost 500 million years old, part of the Ellenburger group of Ordovician rocks. It is in limestone beds in this group that Longhorn Cavern has been formed over many millennia by the dissolving action of downward percolating rainwater. At the top of Backbone Ridge is the side road to Longhorn Cavern. Around the stone administration building are nice surface exposures of dark grey, Ellenburger group limestone.

Longhorn Cavern features dripstone deposits, flowstones, calcite crystals, and solution domes and channels. The cave has been inhabited by humans for perhaps thousands of years. More recently, the

Hoover point roadcut. Note fault offsets in Cambrian beds.

caverns have been used by Indians, outlaws, and by Confederate soldiers who manufactured and stored gunpowder in the cave during the Civil War.

Many of the geologic features in Longhorn Cavern are discussed in the section "Texas Caves and Caverns." For an excellent geological and historical presentation, the booklet *The Geologic Story of Longhorn Cavern*, by W. H. Matthews, III, 1963, Guidebook 4, Bureau of Economic Geology, The University of Texas, Austin, is recommended. It is sold at the Cavern gift shop.

After visiting Longhorn Cavern, return to Park Road 4, turn left (west) and retrace the winding route down the flank of Backbone Ridge. At the bottom of the hill turn left (south) on F.M. 2342 which heads southwest toward Kingsland. This road travels on pink, Town Mountain granite as it parallels the Colorado River and the flank of Backbone Ridge. In four miles where F.M. 2342 makes a T-junction with F.M. 1431, turn left (southwest) toward Marble Falls. In a short distance, the road makes a steep climb to a scenic overlook at the southwest end of Backbone Ridge, where a parking area provides great views up and down Lake Lyndon B. Johnson reservoir on the Colorado River. Across the river to the west is the hilly, granite and schist terrain of the ancient rocks of the Llano uplift; the roadcut at the overlook is a full geologic lesson in itself.

Light-colored cross beds of trilobite hash in green, glauconitic Cambrian sandstone. Hoover Point Lookout, FM 1431.

Marble Falls limestone (Pennsylvanian—300 million years old) in cliffs along the Colorado River below the town of Marble Falls.

In the roadcut, Cambrian rocks, about 500 million years old, display a variety of sedimentary and structural features. As you walk up and down the roadcut, notice the well-defined faults that offset bedding in several places. At the south end of the cut, you can see tilted limestone abruptly faulted down against sandstone. The organic richness of the dark black to green sandstones and shales is immediately evident. Pieces of trilobites, those ancient jointed-leg, segmented creatures that look like modern pill-bugs found in your garden, are common in these rocks. The green color of the sandstone is due to the mineral glauconite, an iron-rich mineral deposited in shallow marine waters commonly as fecal pellets of ocean organisms. The trilobites and glauconite thus place the deposition of this Cambrian sand in shallow marine seas.

Whitish crossbeds are also noticeable in the midst of the dark sands. The crossbeds represent periods when trilobites moulted, and the moults were swept along by currents as particles to form crossbedded "sand," which was then cemented into hard rock by calcite mineralization.

From the Hoover Point Lookout, the road heads down the south flank of Backbone Ridge, and once again traverses the low topography of the Town Mountain granite. Near Marble Falls, granite domes rise above the general plain, and soon giant quarries, their towering cranes and piles of cubic quarried blocks appear near the road.

Granite Mountain has been quarried here since the 1880's, when granite blocks were cut and shipped via a specially constructed railroad line to Austin to build the State Capitol Building. Marketed as "Texas Pink Granite," stones from this quarry grace many buildings throughout the United States and as far away as Iceland and Singapore. Blocks of granite from these quarries were also used to construct the stone groins which extend seaward from the seawall along the coast at Galveston. Across the highway from the Granite Mountain quarries is a roadside park with granite slabs for table tops. Access to an area north of the picnic grounds, via the stiles over the fence, is open to the public; unlike Enchanted Rock, it is permissible here to hammer on the granite. Just east of the quarries, heavily vegetated, upended Marble Falls limestone is downthrown against sparsely vegetated Town Mountain granite along a major fault having at least 3,000 feet of displacement. Between the Marble Falls limestone and the center of the town of Marble Falls, the highway rests on topographically low Smithwick shale of lower Pennsylvanian age.

The dark gray limestone seen around Marble Falls is over 300 million years old, lower Pennsylvanian age. The appropriately named Marble Falls limestone, although no true marble is present, forms cliffs in the canyon walls of the Colorado River beneath the U.S. 281 bridge south of town center. The best view is back toward town from the south side of the bridge. These Pennsylvanian rocks, like the Ordovician rocks seen on Backbone Ridge, are blocks caught and preserved between northeast–southwest-trending faults flanking the eastern edge of the Llano uplift.

Eight miles southeast of Marble Falls, and a mile east of Spicewood on Texas 71, highly deformed shale and sandstone of the Ouachita fold belt peek out from the eroded edge of the Cretaceous rock cover. These folded rocks are mostly covered by Lake Travis, except during periods of drought when they can be viewed in the lower Colorado River Authority Park.

Pedernales River Falls and pools formed on inclined beds of Marble Falls limestone (Pennsylvanian - 300 million years old), Pedernales Falls State Park.

Farm Road 2766
Johnson City—Pedernales
Falls State Park
12 miles

The highway skirts the edge of an erosional window through lower Cretaceous Glen Rose limestone and Hensell sand, where rocks of Ordovician and Pennsylvanian age are well exposed north of, and crossed by, the highway (see map). Generally, the roadcuts are quite poor, though yellow-tan, flat-lying limestone beds in the Glen Rose are present in high areas. Fieldstone along low places bordering the highway are gray, weathered limestone indicative of underlying Ordovician limestone.

The entry road into Pedernales State Park is on yellow-tan lower Cretaceous limestone all the way to the parking area.

The climb down to Pedernales Falls reveals a whole new world, however, as Paleozoic limestone of Pennsylvanian age, about 300 million years old, forms the lower bedrock in the cliffs and falls. The

176

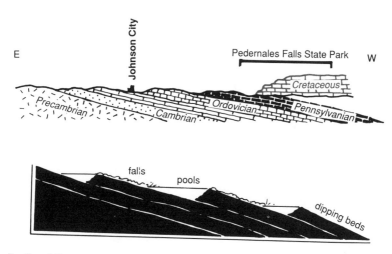

Inclined limestone creates pools and falls at Pedernales Falls State Park.

limestone is very hard and nearly black, and has abundant white calcite veins and large fossil crinoid stems. This is the same kind of Marble Falls limestone that forms the cliffs in the Colorado River Canyon below the town of Marble Falls to the north, where obviously the unit was named. Note how the bedding is inclined, as opposed to the flat-lying Cretaceous limestones above. The Marble Falls limestone was deformed in the Ouachita Mountain-building episode not long after it was deposited near the end of Paleozoic time.

The falls (really rapids) are the product of this inclination. Note how the water slides down the surface of the inclined bedding planes, and how the uptipped ends of these beds form the edges of the pools.

The erosive power of running water is everywhere evident on the limestone surfaces, where grooves, and rills and polished rocks tell of the water's path. Look also for potholes on the edges of the pools. These holes, one foot or larger across, are round, have vertical sides, and usually contain one or more rounded, hard rocks at the bottom. The potholes are ground out of the hard limestone by the swirling action

Cross section of pothole with grindstones in bottom.

177

of the passing water, which moves the round cobbles, causing them to act like grindstones, carving the pothole deeper and deeper. You can even see the grooved sides on some potholes, which record the past grinding action of the trapped cobbles.

White steaks of calcite veins show up in marked contrast to the dark, surrounding limestone. Watch also for white calcitic fossils of crinoid stems, especially on surfaces in the lower part of the falls area.

Fossil crinoid stem in Marble Falls limestone, Pedernales Falls State Park. The penny is used for a scale.

Hill Country Drives
Kerrville—Bandera—Lost Maples and Garner State Parks

Deep, enfolding canyons, pattering streams, high mesas and curving cliff walls of limestone upon limestone are the brush strokes in the geologic painting of this intriguing terrain found southwest of Kerrville and west of Boerne. Many back-country paved roads wind through canyons along streams here, offering the traveler a different pace from the pell-mell freeway rush.

All the rocks in this area are early Cretaceous in age, deposited over millions of years in warm, shallow seas that once covered Texas. The Glen Rose formation, a collection of limestone, shale, marl, and siltstone beds, was deposited along the shifting margins of the sea where dinosaurs roamed in great numbers, leaving their footprints in the sands. The Cretaceous sea then spread over Texas in earnest, depositing thick layers of solid marine limestone, called the Edwards limestone, over the Glen Rose beds. This sequence of strata, Glen Rose below, Edwards above, is found throughout this area.

But, the whole region was uplifted several thousand feet about ten million years ago (Miocene time) which caused streams to erode intensely and cut back the margins of the block. Evidence of the deep stream etching can be seen on the geologic map, where the intricate, feather-like patterns of the Edwards limestone outlines hill tops adjacent to drainages, while the underlying Glen Rose is exposed in the etched-out river bottoms.

Most of the roads in the area follow low areas along stream courses, and in roadcuts, the yellowish, thin-bedded, marl and limestone beds of the Glen Rose can be seen.

179

Geologic map of Hill Country Drives.

Talus on edges of limestone hills.

Eroded material from the mesas commonly forms talus piles on the lower slopes, through which some roads are cut. The chaotic mix of sand, gravel and boulders is a sure indicator of these talus deposits.

In many places in the streams and rivers, low, stair-step waterfalls are found. This phenomenon is a direct result of the bedding character of the Glen Rose formation wherein hard limestone beds alternate with softer marl or silt beds.

Where roads climb over ridge crests, the thick, gray beds of solid Edwards limestone show up in high-level roadcuts. Southwest of Kerrville, Texas 39, F.M. 187 and U.S. 83 top out on the flat, high terrain of the Edwards Plateau. This is a good place to get a feel for the nature of the plateau prior to erosion and where one can also get an appreciation for just how much rock has been removed from the area by stream erosion!

Lower Cretaceous limestone walls in Lost Maples State Park.

Hill top roadcut in Edwards limestone, looking across Hill Country at lower elevations where Glen Rose limestone predominates.

A few common lower Cretaceous fossils found in Lost Maples State Park.

Lost Maples State Park and Garner State Park are located along streams and nestle amongst pleasant canyons and cliffs. Hill tops in the parks are capped by Edwards limestone, whereas Glen Rose rocks are seen low on canyon walls and in the river bottoms. North from Lost Maples Park, F.M. 187 takes you up through the entire Glen Rose—Edwards sequence, and excellent roadcuts are found along this ascent.

This is country made to enjoy at a leisurely pace, and exploring around the canyons and streams and hiking the hills can bring ample rewards of personal geologic discovery. As they say in Australia, "Have a go!"

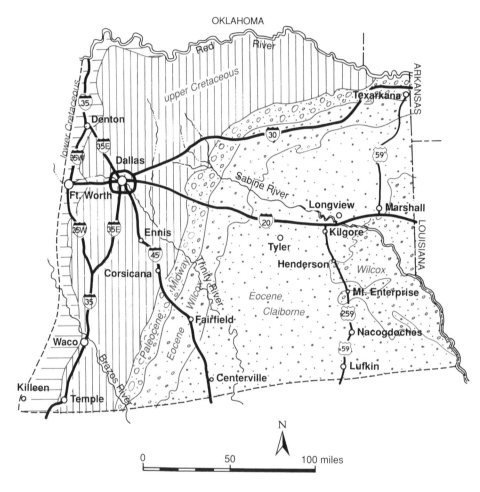

Geologic map of Northeast Texas.

IV
Northeast Texas
Lignite and Piney Woods

Northeast Texas is rich in geological resources—oil and gas, lignite, iron, sand, and gravel—but not much of the geology is seen at the surface. Outcrops and roadcuts are rare, partly because vegetation and thick soils cover the rocks, but more importantly the geologic history of the area for the last 60 million years has been one of nearly continual sinking, with little uplift. Rocks have been buried, not exposed! The geologic story of Northeast Texas is one of tremendous sedimentation and progressive construction of the southern continental margin of North America since the Gulf of Mexico wrenched open as the pull-apart gap between North and South America nearly 200 million years ago.

The geologic map shows a distinctive curved band of parallel packages of sedimentary rocks, each band representing younger and younger rocks toward the southeast.

The western band of Cretaceous-aged limestones which arcs eastward near Dallas toward Texarkana is the approximate location and shape of the southern edge of North America about the time the Gulf of Mexico began to open. The pull-apart, not coincidentally, follows the curve of the old Ouachita Mountain chain. Though North and South America were glued together by collision along the line of the Ouachita Mountains, the Ouachitas were pushed up by this collision to form the supercontinent Pangaea about 300 million years ago. But the weld was evidently not a permanent one, because when it came time for

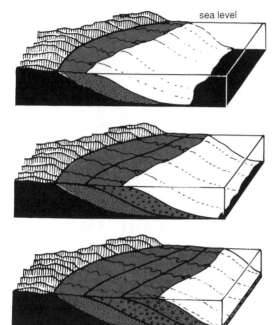

sea level

Progressive filling of the Gulf of Mexico during Tertiary, as thick wedges of sediment built seaward.

Pangaea to break apart in the latest episode of continental drift and sea floor spreading, North and South America split along the zone of crustal weakness defined by the Ouachita Range. The line where the two continents once came together also became the line where they eventually came apart.

The Gulf of Mexico began as a shallow salt pan when the crust sagged and North and South America began to separate about 200 million years ago in the early Jurassic. It wasn't connected to the other oceans at the time, and a thick section of salt, the Louann salt, filled the closed depression. As pull-apart continued, Gulf waters opened to the other oceans, and by early Cretaceous time, 140 million years ago, shallow marine seas covered the Gulf and lapped over the edge of the continental margin of North America. Limestones, built of the shells of marine organisms living in the shallow, early Cretaceous seas began to accumulate. Such rocks are seen around Ft. Worth. The Gulf continued to sag, and by late Cretaceous time, 90 million years ago, the shoreline had built farther into the Gulf, but more rocks and limestones were deposited further offshore in the Gulf. These rocks, of late Cretaceous age, are seen in the Dallas area.

By 60 million years ago, at the end of the Mesozoic era, highlands rose in the western interior of North America and the sinking rate of the Gulf must have increased. Tremendously thick piles of sediment were dumped into the Gulf by ancestors of the Sabine, Trinity, Brazos, and Colorado rivers. A progressive sequence of sediment wedges, composed of river channel sands, muds, and peat deposits, and sandy shorelines were added onto the North American continent from 60 million years ago, until today. The process still goes on. This progressively younger and younger pile of sandy and muddy sediments is seen southeast of Dallas — Ft. Worth.

Lignite, which is compressed peat derived from plant material, is mined extensively along a wide band in northeast Texas. This Eocene deposit, 50 million years old, was laid down in the muddy areas between river channels, where plants grew abundantly.

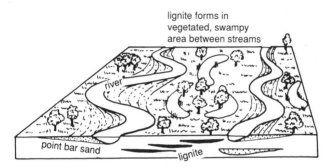

Depositional setting of lignite. Plant debris is pressed into peat; when buried it becomes lignite, which turns into coal if pressured enough.

With such rich organic sediments, as well as closely associated sands, it is no wonder northeast Texas is also rich in oil and gas deposits. Organic Cretaceous limestones, as well as the organic Eocene shales are the sources for the hydrocarbons, while associated porous sandstones and limestones form the reservoirs into which the oil and gas migrated. These rocks were buried deeply where they could be cooked at the right temperatures to generate oil and gas.

At the surface, the sandy early Tertiary rocks form an ideal substrate for pine trees, and in the eastern half of northeast Texas, tall stands of pines predominate in the Sandy Hills region.

With this story in mind, look carefully for the rare rock exposures you perchance come across as you travel the roads of northeast Texas.

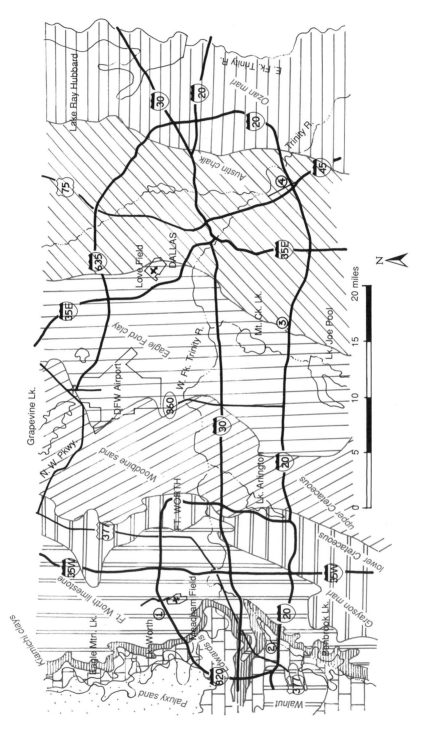

Geologic map of Dallas—Ft. Worth.

Austin chalk on Interstate 20 south of Dallas at Exit 474 on Texas 310 just north and east of Interstate 20 and I-45 intersection.

Interstate 20 & 635
South of Dallas—Ft. Worth
65 miles

Interstate 20 south of Ft. Worth and Dallas makes a traverse across bands of Cretaceous rocks. The youngest band is in east Dallas, the oldest west of Ft. Worth.

Several excellent roadcuts along the freeway are plotted on the Dallas—Ft. Worth geologic map, though for most of the drive you won't see many rock exposures.

Location 4 on the map is a wonderful roadcut in the Austin chalk. Take exit 474 off I-20, onto Texas 310 going north. It is the first road east of the I-20 and I-45 intersection.

The 310 roadcut is in white, marly limestone, typically Austin chalk, and is full of fossils. Clams and oysters are most common, but pieces of huge clams up to several feet across are very common, and very characteristic for this time period in the upper Cretaceous. They appear as flat dinner-plates, about 1/4 inch thick. The clam built its shell of calcite crystals arranged like stand-up toothpicks, perpendicular to the shell surface (see the photo).

Shells of this clam, (called *Inoceramus*) are so big you probably won't find a whole one!

Clam shell, Inoceramus

Calcite has replaced the original clam shell material.

Trinity Trail on the Clear Fork of the Trinity River. South-west of Ft. Worth on the Southwest Loop at the Ridglea Country Club Drive exit.

Lower Cretaceous Goodland limestone.

West of the I-45 crossover, I-20 rides along on white limestone terrain of the Austin chalk. You will see a few exposures along the way, until five to seven miles west of I-45, the road descends into the Mountain Creek valley (exit 458). On the south side of the highway, (location 3 on the map) the white limestone beds of Austin chalk are exposed on the ridge top, above a gray section of Eagle Ford shale. The chalk plainly forms the ridge crest, whereas the soft shales of the Eagle Ford form the valley where Mountain Creek found it easy to cut its way downward.

Roadcut in Austin chalk over Eagle Ford shale on Interstate 20 at Exit 458 southwest of Ft. Worth, looking south.

Farther west, Highway 183 (SW Loop) leaves I-20, and soon crosses the Clear Fork of the Trinity River. Exit at Ridglea Country Club Drive and turn back under the SW Loop, which brings you to river side and the Trinity Trail along the Clear Fork. Here you will see good river bank exposures of light tan, lower Cretaceous beds of Edwards limestone, full of fossil snails, clams, and oysters. This locality is marked 2 on the Dallas—Ft. Worth geologic map.

Inoceramus

Interstate 20
Dallas—Marshall
132 miles

The pine-covered, Sandy Hills of northeast Texas follow the outcrop band of Eocene Claiborne sandstones. The eastern two thirds of I-20 between Marshall and Dallas are underlaid by Eocene sedimentary rocks, mainly sandstone and claystone, whereas the western third is on upper Cretaceous limestone country, except for a narrow band of poorly exposed Paleocene Midway rocks. The highway runs perpendicular to these bands as shown on the Northeast section geologic map.

Outcrops and roadcuts are not common, although you will see a few pretty good ones where I-20 crosses the Sabine River about 25 miles west of Marshall.

For about 70 miles west of Marshall, Interstate 20 rolls along on the hilly topography of the Sandy Hills, where river drainages have created low hills and valleys as they carve into the soft Eocene sandstone bedrock. Pine trees grow in profusion on the Sandy Hills,

Roadcut in Tertiary sediments west of the Sabine River on the north side of Interstate 20, 25 miles west of Marshall.

Close up of ripples, cross bedding, and iron concretions.

as pines prefer sandy, well-drained soils, with relatively high rainfall. And, the annual precipitation around Marshall is about 48 inches per year, while Dallas gets about 36 inches of rain, on the average.

Near the town of Longview, about 20 miles west of Marshall, look for large cranes north of the highway. These large drag-lines, with their gigantic buckets, are used to mine lignite, which can either be described as hard peat or soft coal. A swath of lignite crosses northeast Texas; where it is close to the surface, it is commonly mined and burned to generate electricity. The plant material that was compressed to form lignite was deposited in floodplains between river channels during Eocene time 50 million years ago.

Watch for the Sabine River crossing about 25 miles west of Marshall. Gray sandstone and shale are seen in river bank exposures, though they are not spectacular by any means. However, there is an excellent roadcut west of the Sabine River on the north side of I-20. This is the best one along the entire stretch between Dallas and Marshall. Here sandstone channels display cross-bedding and ripples, gray claystones are exposed, and beds of bright red to black ironstone concretions are seen. The outcrop is colored yellow and red from limonite (iron oxide) staining. The Claiborne group rocks here were laid down on a floodplain in Eocene time.

At Kilgore, the importance of oil and gas to this region is demonstrated in the excellent East Texas Oil Museum. And, to further illustrate this importance, don't miss the oil derrick picnic tables at the rest area east of Starrville!

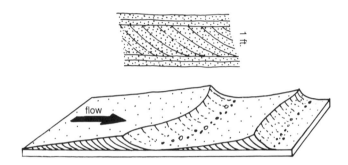

Wave-like bed forms produce cross-bedding, where water moves the sand in linear ridges along the bottom, pushed by currents or waves.

In the central segment of I-20, small roadside exposures seem more common. North of Tyler, just east of Exit 567, gray sandstone, claystone, and a black lignite bed appear on the south side of the highway. Not far away at FM 5015 exit similar rocks are exposed on both sides of the highway. Watch for more such roadcuts in this area—they are small, but are windows into the bedrock of Eocene rocks that form the substrate in northeast Texas.

Interstate 20 crosses the mainly flat, agricultural Black Prairie region east of Dallas. Pine trees don't grow here, and rocks are nowhere to be seen, except for white, limestone rubble here and there in fields, where Cretaceous limestones form the underlying bedrock of this area.

Interstate 30
Dallas—Texarkana
170 miles

The western third of I-30 between Dallas and Texarkana traverses a band of Cretaceous limy rocks in the Black Prairie area, while the eastern two-thirds crosses soft, sandy-clayey sediments of early Tertiary (Paleocene and Eocene) age. But, you will see more on the geologic map than along the highway, because few roadcuts or outcrops are to be seen anywhere along the way. Light-colored limestone rubble in fields near Dallas, and gray clay in stream banks farther east, are about the only roadside indicators of the geology.

Notice on the geologic map how the bands of Paleocene and Eocene rocks swing eastward toward Texarkana. This curve represents the position of the edge of the early Gulf of Mexico shortly after the Gulf opened up as North America split from South America. Since that time, wedges of sand and clay sediment have been progressively added to fill in the Gulf of Mexico, until today's shoreline is at Galveston, about 225 miles away.

In other words, in the last 60 million years since Paleocene time, 225 miles of land have been eroded from the North American continent and dumped into the Gulf of Mexico by the erosive and transporting power of streams. But, this sediment wedge is not just 225 miles wide and one foot deep either. Some geologists estimate this Tertiary pile of sediment to be 40,000 feet thick!

Six miles east of the I-30 and I-635E junction, I-30 crosses Lake Ray Hubbard, an artificial lake created by the damming of the east fork of the Trinity River. The Trinity heads in the Rolling Plains and western Cross Timbers country northwest of Dallas—Ft. Worth, and becomes one of East Texas' major streams, ultimately flowing into Galveston Bay east of Houston.

A mile east of Greenville, the freeway crosses the Sabine River, barely a creek here. The Sabine flows southeastward from headwaters east of Dallas in the Black Prairie, to become a sizeable river marking the state border with Louisiana. It empties into the Gulf of Mexico at Port Arthur.

Near the town of Winfield, just west of Mt. Pleasant, watch carefully for large earth piles, crane-like drag lines, and a mine headframe. Here, 50 million year old, Eocene lignite beds are being mined in open pits. The lignite is black, highly compressed plant material on its way to being coal. The progression is: plants–peat–

Lignite mine north of Winfield, Texas on Interstate 30.

lignite–bituminous coal–anthracite coal. The plant material was deposited around the early edge of the Gulf of Mexico in swampy areas between river drainages. Similar environments exist today landward of the Texas barrier islands, and the drive from Houston to Galveston emulates a trip through the environment of deposition of this 50 million year old lignite.

The mining process requires the removal of overburden, which is stacked in large piles, the large earth piles you see from I-30. Then the lignite is excavated by huge buckets and drag lines, the crane-like devices. As the front face of the mining moves forward, the overburden is continually filled in behind, graded to natural contour and planted, so that only a narrow strip of active mine is open at any one time. This new method of open-pit mining is more environmentally accepted and is cost effective.

The lignite is used for fuel in power plants that generate electricity for Dallas–Ft. Worth.

In river banks where the freeway crosses White Oak Creek and the Sulfur River, 20 and 25 miles east of Winfield, gray clays of Eocene age are exposed. But, look carefully, the outcrops are small.

Between Mt. Pleasant and Texarkana, notice how pine trees become more common eastward. Several factors contribute to pine growth here. First, the soils are sandier because sandstone is more common in this part of the Eocene Wilcox formation, which forms the substrate for the soils. Pines prefer well-drained sandy soils. Second, rainfall increases from 36 inches per year measured at Dallas, to 48 inches per year at Texarkana, and pines require this added moisture.

The countryside becomes a little hillier near Texarkana, as the road crosses the topographic relief created by lateral drainages of the Sulfur and Red rivers.

Interstate 35E
Denton—Dallas
28 miles

This freeway segment travels mainly through urban areas. The bedrock is on upper Cretaceous Eagle Ford and Woodbine sandstone, shale, and limestone strata, though exposed rocks are rarely seen from the expressway. One good outcrop of Woodbine sandstone does occur, however, northwest of the bridge over Lewisville Lake. A nice section of red soil and white, rounded sandstone can be seen along the lakeshore. This sandstone was probably deposited along the shoreline of the late Cretaceous Seaway.

Interstate 35 & 35W
Ft. Worth—Waco—Temple
115 miles

From the junction of I-20 and I-35W southwest of Ft. Worth to Waco, the freeway travels along the flat terrain of the Grand Prairie. Underlying bedrock is lower Cretaceous limestone, though none shows up along the highway. In the distance to the east is a prominent tree-covered topographic ridge known as the Eastern Cross Timbers.

The ridge swings closer to the highway near Alvarado, and geologic mapping shows the bedrock holding up the ridge to be upper Cretaceous Austin chalk. The soft, upper Cretaceous Eagle Ford shales, here below the road, have been eroded by streams, and the highway traverses this low, flat area. Rocks are not easily seen along this stretch of road.

Near Waco, the hills to the west are on the edge of the Comanche Plateau, an uplifted area where Cretaceous limestones are the near-surface bedrock. The highway crosses the Brazos River in the center of Waco. The Brazos ("arms" in Spanish, originally called "Brazos de Dios," or Arms of God), nearly bisects Texas from northwest to southeast.

Tributaries of the Brazos have their headwaters on the High Plains Plateau in the Texas Panhandle. The Brazos frequently flows red-brown from the red-colored sediment it picks up as it winds across Permian red beds in the country north of Abilene.

From Waco, the Brazos River continues flowing southeastward, paralleling Texas Highway 6 past Bryan, College Station, Washington (where the first capital of Texas—"Washington-on-the-Brazos" is located), Richmond-Rosenberg, and finally Freeport, where it empties into the Gulf of Mexico. The Brazos River was considered by Stephen F. Austin to be the future major waterway for trade across Texas when he established a colony at Ft. Bend in 1821. Paddle-wheel boats did move up and down the Brazos carrying cotton and trade goods for many years until the advent of railroads and dams.

Between Waco and Temple, the eastern edge of the uplifted Lampasas Cut Plains is visible to the west as a low line of skyline hills. The road lies on Austin chalk bedrock, though none is seen except in fleeting glimpses of small exposures beneath highway bridges and stream banks.

Temple calls itself the "Wildflower Capital of Texas." This is claiming a lot in a state where wildflowers are highly prized and widely planted, especially along public roads.

Interstate 45
Dallas—Corsicana—Centerville
110 miles

Bands of upper Cretaceous marlstone, limestone, and shale underlie the Interstate between Dallas and Corsicana, though the landscape is fairly flat and rocks are not commonly seen along this stretch.

At Ennis, the road swings east, then south again in a few miles, toward Corsicana. The Corsicana oil field, located near the southward bend in the Interstate is where the first major oil in Texas was discovered in 1894. The town of Corsicana marks the edge of the upper Cretaceous belt and approximates the position of the original edge of the Gulf of Mexico coastal plain which began to receive huge amounts of sand, silt, and clay sediments about 60 million years ago, a process that continues today.

South of Corsicana, the freeway traverses a series of sand, silt, and clay bands that represent successive episodes of Gulf-filling. Sandy ridges south of Corsicana are Paleocene, the oldest in the sequence, becoming younger and younger southward toward Houston. Off to the west are low skyline ridges held up by Cretaceous rocks. Cretaceous rocks are separated from the younger Tertiary rocks by a series of linear, north-south faults that are related to the timing and trend of the Balcones fault system farther south. Here, the so-called Mexia

199

fault system was also activated in Miocene time, ten million years ago, and like the Balcones fault near Austin, this one follows the edge of the old buried Ouachita mountain range trend.

Low rolling hills characterize the countryside between Corsicana and Fairfield, indicating sand and shale beds underlie this stretch. South of Fairfield, near Buffalo, a rather prominent ridge crosses the Interstate south of town. The redness of the soil and few sandy exposures are typical of the Eocene Carrizo sand, which forms a distinctive ridge southwestward past Austin.

Near Fairfield are four major salt domes which have pushed older Cretaceous caprocks to the surface to nestle amongst the surrounding Tertiary terrain. Upper Gulf Coast salt domes are numerous in the subsurface around this area, and the four surface domes are merely an expression of this extensive array of subterranean salt upheavals.

From Buffalo south to Huntsville, the road rolls up and down as successive sand ridges are encountered. A prominent Eocene Sparta sand ridge is located about ten miles south of Buffalo. Here the freeway climbs out of Keechi Creek and over the Sparta sand ridge, though no obvious outcrops or roadcuts are to be seen.

Centerville marks the end of this section of road. The Centerville to Houston segment is described in the Southeast section.

Salt domes and Carrizo sandstone in the Fairfield area.

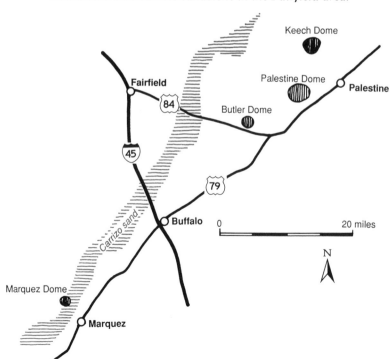

Limestone quarry of the Cretaceous Kiamichi formation on I-820 west of Ft. Worth.

Interstate 820 & 35 W
Ft. Worth—Gainesville—Red River
63 miles

From the junction of Interstate 820 with Interstate 30 on the west side of Ft. Worth, northward along I-35W and I-35 to Denton, Gainesville, and the Red River, the Interstate travels along the flat terrain of the Grand Prairie. Rich soils and farmland spread outward on either side of the road, but few outcrops or roadcuts are to be seen along the way. This part of the Grand Prairie is underlaid by a belt of flat-lying, lower Cretaceous limestone, sandstone, and shale beds that were deposited near the shoreline of the inland Cretaceous Seaway. As this sea advanced across middle America, shorelines pushed outward over clear marine water, laying sandstone and mudstone over limestone. When sea level rose, marine limestone was laid back over the retreating sandy-muddy shoreline. This pattern repeated many times; alternations of sandstone, limestone, and mudstone (shale) are common in this interval of the Cretaceous.

As I-820 leaves the I-30 intersection, note that the highway is built on a high spur, which is held up by Cretaceous limestone, though no rocks are visible. As I-820 crosses the bridge over Lake Worth, look

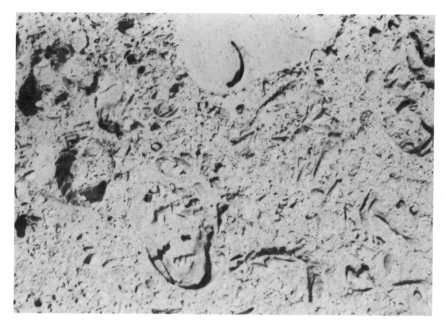

Sign at Tourist Bureau made of Cretaceous limestone north of Gainesville on Interstate 35.

east to see a nicely exposed bank of lower Cretaceous sandstone/limestone at water level. Note the beds are flat-lying.

About nine miles from I-30, watch for a huge quarry on both sides of I-820, just south of the Meacham Field turnoff. Again, flat-lying beds of lower Cretaceous gray shale, limestone, and marl of the Kiamichi formation are beautifully exposed, see (1) on map on page 188.

About one mile north of I-820 junction, I-35W crosses Fossil Creek, where east of the road you may get a fleeting glimpse of lower Cretaceous limestone and clay beds in the creek banks.

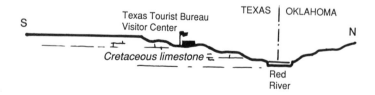

From here northward, I-35W and I-35 is on flat terrain all the way to Gainesville. However, north of Gainesville, more topography is encountered as the expressway nears the Red River. Side drainages have cut into the bedrock here, and a few exposures of light-colored limestone are seen near the picnic area on the east and the Texas Tourist Bureau building on the west. The steep river bluffs cut into Cretaceous limestone, but none can be readily seen from the bridge.

The Red River is aptly named, for it often runs bright red! The headwaters of the river are on the high plains of the Texas Panhandle, where the Red River's tributaries have cut canyons through bright red Permian shales and sandstones. The river continues to run eastward across the Rolling Plains where more red Permian rocks are exposed. The red sediment carried by the river gives the Red its characteristic color.

Sand bars on the Red River north of Gainesville on Interstate 35. Oklahoma is to the left and Texas to the right.

U.S. 59
Texarkana (I-30)—Marshall (I-20)
68 miles

The road between Texarkana and Marshall is a lovely drive through dense pine forests, small Texas towns, and up and down amongst sandy hills. Between Texarkana and Lake Texarkana, the bedrock is Eocene Wilcox sandstone and shale. From Lake Texarkana southward, the towns of Atlanta, Linden, Jefferson, and Marshall reside on younger Eocene Claiborne sandstones and shales. The highway traverses the topographic area of East Texas known as the Sandy Hills. The reason for the hilly relief is that bedrock is cut deeply by side drainages that run perpendicular to the road as they make their way toward the Red River only a few miles to the east. The road simply goes up and down across one drainage and drainage divide into another and then another.

Watch for an excavation wall behind the Pizza Hut in Atlanta, where you can see a nice section of red soil over Eocene sandstone and claystone. Halfway between Linden and Jefferson is a tan sandstone roadcut on the east side of the highway across from a Shell gas station. The claystones were deposited in flood areas between Eocene river channels, while the sandstones were laid down as sandy beds in the river channels.

U.S. 59 - 259
Lufkin—Nacogdoches—Kilgore
79 miles

This segment of U.S. 59 crosses tilted Tertiary sandstone and shale beds, youngest to oldest from south to north. These rocks are part of the wedge-shaped package of sediment that has been filling the Gulf of Mexico since the time of dinosaur demise, or for about the last 60 million years of earth history. Don't expect to see rocks jacked up at a noticeable angle in roadcuts, though, because the angle of tilting into the Gulf is about one degree! This may not seem like much, but if rocks which are at the surface near Nacogdoches tip toward the Gulf at only one degree, they would be buried over 2½ miles deep, or nearly 14,000 feet beneath Galveston!

Between Lufkin and Nacogdoches, the highway crosses the Angelina River which is the main feeder stream to the nearby Sam Rayburn Reservoir. Two miles south of Nacogdoches, opposite the entrance to the Piney Woods Country Club, is a quarry cut of orange-tan sandstone of the Sparta sandstone. The Sparta is generally a ridge former, and hilly topography can be traced across the countryside where the Sparta comes to the surface.

In Nacogdoches watch for a very good roadside cut on Business 59 at Smith Street, where another Eocene unit (Weches formation) can be viewed up close. Here red, weathered siltstones and sandstones overlie green, unweathered siltstones and sandstones. Careful examination will show extensive churning by marine organisms. Marine

Red (weathered) above green (glauconite-filled) Eocene Weches formation claystones at Nacogdoches, Texas on US 59.

Beautiful, tabular cross beds in Sparta sandstone, along US 59 north of Nacogdoches, Texas.

fossils are abundant not only in this roadcut, but in many other rock exposures in and around Nacogdoches. The green color is from the iron-rich mineral glauconite, which forms in shallow marine shelf waters. As the iron in these rocks weathers, it oxidizes, or 'rusts' and turns red. Nacogdoches is home to Stephen F. Austin State University, and claims to be the 'oldest town in Texas.' The old stone fort near the college was built in 1779.

The band width of the Sparta sand extends north from Nacogdoches for a few miles, and a few roadcuts of the orange-tan sand are seen near the U.S. 59–259 junction. This guide follows 259 to Kilgore.

Seven miles north of the U.S. 259–59 junction watch for an excellent roadcut in Eocene sandstone where beautifully preserved cross-bedding, with iron-staining which accentuates the beds, can be

Gray shale and black lignite beds of the Eocene Wilcox formation 7 miles north of Mt. Enterprise, Texas on US 259.

seen. The straightness of the cross-beds indicates they were deposited in rather slow-flowing and not very turbulent water. The red color from weathered glauconite further indicates deposition in shallow marine water.

Around the town of Mount Enterprise, notice the hilliness and bright red soils and occasional roadcuts, as the highway crosses the colorful outcrop band of the Eocene Carrizo sand. There is a red Carrizo band south of Mount Enterprise, then another one north of town. The Carrizo is actually repeated here because of the east-west-oriented Mount Enterprise fault which crosses U.S. 259 just south of town. The north side of the fault is the down-dropped side. The cross-section diagram shows how the two Carrizo sand belts on either side of Mount Enterprise are caused by faulting.

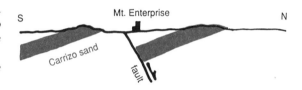

Cross-section showing how two Carrizo sand bands on either side of Mt. Enterprise are caused by faulting.

The roadcuts get better toward Henderson. About seven miles north of Mount Enterprise, be on the lookout for a gorgeous roadcut on the east side of the highway, where striking black and gray alternating beds of lignite and shale stand out in bold relief. The papery lignite and thin beds of gray, intervening shale, containing ironstone concretions, were deposited in a swampy floodplain between river channels. The rich cover of plants contributed their debris to form the black, carbon-rich lignite layers.

Rolling countryside speaks of alternating sandy and shaly sequences of rocks, where the sands form ridges and soft shales erode to form valleys.

At the junction of U.S. 79 and 259 in the town of Henderson, bright red, orange and tan Eocene Carrizo sandstone exposures form a backdrop to the buildings along the roadway. The darker reddish portions are sand channels, whereas the lighter, nearly white beds are clay layers deposited as flood deposits next to the original Eocene channels.

*Red Eocene
Carrizo
sandstone in
Henderson,
Texas at US
79-259
junction.*

From Henderson to Kilgore the topography continues to roll along—more sand and shale alternations, the Carrizo sand extending northward about half the distance to Kilgore.

At Kilgore is Kilgore College, and the marvelous East Texas Oil Museum, located in town right on U.S. 259. The museum offers a historical look at a full-size street scene of a 1930's oil boom town, complete with muddy street, shops, movie theatre, post office, and barbershop where you can hear all the news of the day. Murals,

The East Texas oil field.

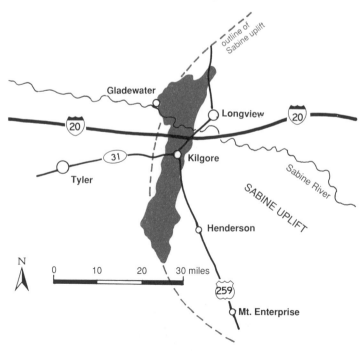

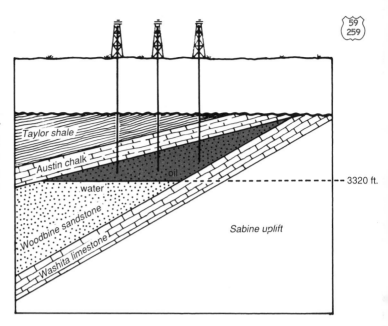

Cross section of East Texas oil field, showing key elements of this giant stratigraphic trap.

Taylor shale

Austin chalk

oil

water

Woodbine sandstone

Washita limestone

Sabine uplift

3320 ft.

displays, original equipment and tools tell the exciting story of discovery of the gigantic East Texas oil field by 'Dad' Joiner in 1930. The field virtually engulfs the town of Kilgore (see map) and is the largest oil field in the United States at five billion producible barrels, second only to Prudhoe Bay in Alaska which is twice as big. The field still produces oil from the pinch-out edge of the Eocene Woodbine sand. The museum charges a small entry fee, but it is one of the best small museums in Texas, and well worth the visit.

East Texas Oil Museum, Kilgore, Texas.

EAST TEXAS OIL MUSEUM

Big feldspar crystals. Town Mountain granite on Enchanted Rock.

The Enchanted Rock name comes from old Indian legends and pioneer observations of strange sounds and lights. Common creaking and groaning noises could easily be granite blocks grinding against one another as they expand and contract from heating and cooling between daytime and nighttime. Low light at dawn and sunset shimmers off crystals in the granite and the sparkling reddish domes blend almost magically with the early morning or late evening sky. Thus, geology explains legend, but should not reduce our sense of wonder and awe over natural phenomena.

The dome shape of the hills is the product of interaction of erosion with the geologic peculiarities of the granite. Note on the sketch map how the domes are separated by linear zones of fractures that intersect at nearly right angles. Search around in the low area between Enchanted Rock and Little Rock along the Echo Canyon Trail to find areas where blocky fractures show up quite nicely. Water and erosion attack these fractured areas with greater ease than the solid granite areas, creating valleys along the fractures, while leaving solid granite to form the high areas.

The rounded shape of the domes is caused by exfoliation. Granite forms deep within the earth's crust; as thousands of feet of rock overburden are removed by erosion over geologic time, pressure caused by the weight of this pile of rock is reduced. The granite expands a little in response to the lessened pressure, which in turn causes the granite to split in curved sheets. Weathering cracks the

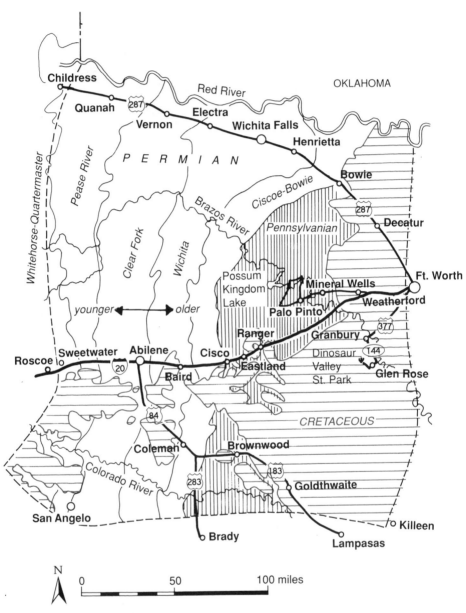

Geologic map of Northcentral Texas.

V
Northcentral Texas
Red Beds and Dinosaur Tracks

Eye-dazzling red beds in canyon walls, stream banks, and roadcuts are the traveler's treat in the west half of northcentral Texas. Rivers flowing through this land take on the eroded sediments' hue, so much so that one great river, The Red, is simply, but perfectly named. In the east half the red beds are gone, but dinosaur tracks on pale colored rocks in stream beds easily make up for the color loss! Badlands, canyons, colorful rocks, high ridges, fossils, and tracks are to be found in northcentral Texas, though these scenes are commonly separated by long stretches of low-topography farmland.

Three large, distinctive patches of rocks make up northcentral Texas in the area between the Red River on the north, the Colorado River on the South, the High Plains escarpment on the west, and Interstate 35 through Ft. Worth on the east. The geologic map of northcentral Texas clearly shows north-south-oriented, parallel bands of rocks. Looking like a tipped stack of cards, these bands are among the oldest sedimentary rock exposures in the state. Deposited during the late Paleozoic period, the bands are regularly stacked, oldest to youngest, from east to west. Around Mineral Wells and Possum Kingdom Lake, the rocks are Pennsylvanian in age (290 million years old). West of there, around Wichita Falls and southward to Abilene, the rocks are a bit younger, Permian in age (250 million years old).

The erosive power of the Brazos River system, which cuts diagonally across the center of northcentral Texas, has removed an im-

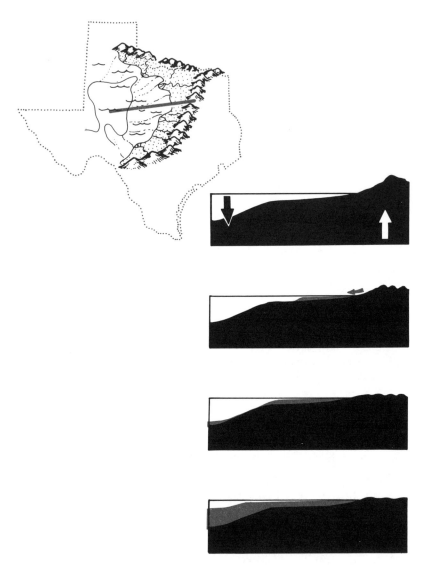

Filling of the Permian Basin with sediments.

pressive amount of overlying rock to uncover these Permian and Pennsylvanian rocks. A high rim of younger Cretaceous rock to the east and south is remnant testimony to the complete cover by Cretaceous sediments, which once extended across northcentral Texas.

The bands of Pennsylvanian and Permian-aged rocks are the result of progressive filling of a deep crustal basin in southwest Texas, created by the sinking of the crust in that area, while the crust rose in central Texas and adjacent Oklahoma. About 300 million years ago, the Wichita, Arbuckle, and Ouachita mountains were pushed upward along the southern margin of North America as it collided with Africa and South America to form the great supercontinent Pangaea. Southwest of this triangle of mountain ranges, the crust collapsed to form a deep basin known as the Permian Basin. A broad shelf-like area, now northcentral Texas, existed between the mountains and the basin for several million years. Across this shelf swept rivers and streams carrying and depositing their eroded sedimentary load etched from the mountains. As the mountains were worn down, the debris from them was deposited on the shelf as river sand and floodplain mud, while red mud, sand, and evaporative gypsum were deposited in the west, where the climate was dry and shallow water quickly evaporated. Even further west, around the margins of the clear ocean waters of the deep basin, thick sections of limestone accumulated during the Permian to form the now magnificent cliffs of Guadalupe Mountains National Park.

But the main story of late Paleozoic time in northcentral Texas is one of sediments filling up a basin! The orderly stack of sediments is over 5,000 feet thick, tilting about ½ degree to the west. Once the basinal area was filled, and the stance of the mountains greatly reduced, a rather flat plain, near sea-level, was all that remained. On the late Paleozoic river banks and floodplains of northcentral Texas a dramatic biologic story unfolded, as early amphibians and reptiles first walked, then strode mightily into the Mesozoic to spawn a great race of reptilian giants—the dinosaurs.

By Jurassic and Cretaceous time (200 to 60 million years ago) the continents danced apart once again, new seas opened up, and sea level rose across the flat plains to deposit a thick, widespread layer of limestone from its marine waters, while sandstone and mudstone were laid down on the coastal margins. Along this seaway mighty beasts shook the ground as they marched across firm, sandy flats. Their tracks are well-preserved in the Cretaceous beds of northcentral Texas, and Dinosaur Valley State Park preserves some of the best.

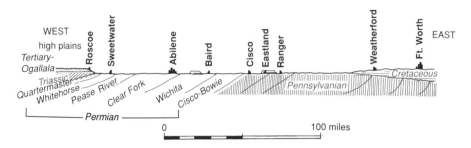

WEST   Roscoe   Sweetwater   Abilene   Baird   Cisco   Eastland   Ranger   Weatherford   Ft. Worth   EAST

high plains

Tertiary-
Ogallala

Triassic

Quartermaster

Whitehorse   Pease River   Clear Fork   Wichita   Cisco-Bowie   Pennsylvanian   Cretaceous

——————— Permian ———————

0   100 miles

Cross section along I-20 between Ft. Worth, Abilene, and Roscoe. The tilt of the rocks is greatly exaggerated.

Interstate 20
Roscoe—Abilene—Ft. Worth
190 miles

This 190 mile stretch of Interstate 20 between Roscoe and Ft. Worth is a perfect transect of the geology of northcentral Texas. The highway runs perpendicular to the colorful red bands of Permian rocks in the western half between Roscoe and Cisco, then cuts across the southern part of the older Pennsylvanian band of rocks between Cisco and the Brazos River crossing, and finally traverses younger, limy Cretaceous rocks between the Brazos River and Ft. Worth. Though rocks are not exposed at the surface over much of this area, the vegetation is frequently a clue to the underlying strata. Mesquite commonly grows over claystones, while post oaks favor sandstone terrain. Junipers and live oaks prefer limestone country, so if you see green trees and shrubs in the winter, you are on calcareous soils and rocks.

Roscoe lies on the caprock of the high plains, and between Roscoe and Sweetwater the Interstate eases downward onto the surface of older Permian rocks. The escarpment east of Roscoe is not very impressive, as it is in other places, but red bed roadcuts a few miles east of Roscoe are tell-tale signs you've crossed over into Permian terrain. Eight miles east of Roscoe, a gypsum plant is located on the south side of the freeway. The gypsum is mined from the Permian Whitehorse formation, where extensive flats acted as evaporation pans during the Permian, laying down bed after bed of drying gypsum along with bright red, oxidized sand and mud layers. Gypsum,

214

calcium sulfate, is used in making "plaster of Paris" and is the white sandwich filling in the wallboard of your house.

Between Sweetwater and Trent, red-bedded sediments of slightly older Permian rocks, the Pease River group, are exposed north of the highway, whereas south of the road along the skyline is a high, flat-topped ridge, the Callahan Divide. This ridge is an erosional remnant of a once-extensive cover of Cretaceous rocks. The road travels on the surface of red Permian rocks - the Cretaceous rocks of the skyline hills rest on top of this surface. The Callahan Divide is a northern outlier of Cretaceous limestone beds that make up the Edwards Plateau and Hill Country in central Texas to the south.

West of Trent is a highway rest stop where you can get good views of the surrounding countryside from the elevated topography. At the Trent exit watch for red, inclined beds of sandstone. This is a pretty good roadcut of Permian San Angelo formation sandstone - part of the Pease River group.

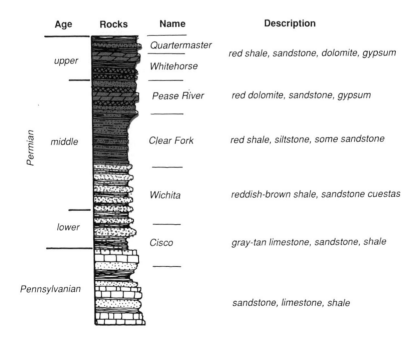

Sequence of Permian rocks in Northcentral Texas.

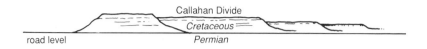

How Cretaceous cover has been eroded by streams to expose underlying Permian rocks, and create ridges and mesas.

East of Trent, low escarpments are seen extending out for many miles on either side of the expressway. These cuestas are formed where slightly harder, west-dipping sandstone beds come to the surface. Erosion works a bit faster on the softer, surrounding mudstones, so the sandstones are left standing higher; hence the cuestas (see diagram). This sandy part of the Permian is in a third, slightly older band of rocks, the Clear Fork group. This band is fairly thick and hence wide; it extends to the east side of Abilene.

The ridge of Cretaceous rocks can still be seen on the skyline south of Abilene. The road between Trent and Abilene is quite flat, as is the surrounding terrain. The smooth, flat ride continues about 15 miles east of Abilene. Though another belt of Permian (Wichita group) rocks has been encountered, it is difficult to distinguish because of the flat terrain and few roadcuts or outcrops.

About 15 miles east of Abilene, near the town of Baird, the road climbs over a spur of Cretaceous rocks and several excellent highway cuts nicely display the internal nature of these rocks. Note the thin beds of limestone, sandy limestone, and mudstone that are tan and gray, distinctively not reddish like the surrounding Permian rocks. The limestones have abundant clams and snails—fossils that indicate deposition in clear, shallow marine water.

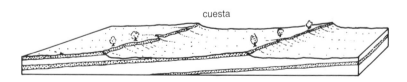

Cuestas—ridges where edges of inclined strata are differentially eroded at the surface, are common along Interstate 20.

216

Fossiliferous, Cretaceous limestone, Interstate 20 roadcut near Baird.

East of Baird, brown sandstones hold up a line of cuestas. These are typical sandstones of the Permian Wichita group of rocks, deposited as sands of meandering streams, which wandered westward across a vast floodplain in Permian time, carrying their eroded loads westward from the Ouachita Mountains.

Between Baird and Cisco, sandy cuestas are commonly seen extending away from the highway on both sides. These sands represent the lowermost Permian Cisco-Bowie deposits of northcentral Texas. The brown sandstones and varicolored shales were deposited in stream beds and interstream areas, respectively, on deltas built along the early Permian shorelines.

The town of Cisco is the demarcation between Permian rocks to the west and Pennsylvanian rocks to the east. Blocks of tan-orange limestones and sandstones are exposed on ridgetops and old roadcuts a few miles east of Cisco. But, between Eastland and Ranger, light tan limestone outcrops and quarries are in Cretaceous terrain—another spur of younger rocks that extends northward from the main body of Cretaceous rocks, and here overlies the older Pennsylvanian section. Around Ranger, the ledge-bordered hills and ridges are built of Cretaceous limestone and sandstone.

Between Ranger and the Brazos River, Interstate 20 crosses a series of prominent cuestas formed by resistant beds of limestone and sandstone. Softer claystone and coal beds are found in the low, eroded areas between the cuestas. Near Thurber, for example, look for the

217

Roadcut of middle Pennsylvanian-Canyon group tan sandstone on Cuesta ridge at the east exit to Ranger, looking north.

smokestack as a marker to identify nearby weathered coal beds. The depositional environment during the late Pennsylvanian was quite interesting because of its cyclic nature and variety of sediments, which led to economic accumulations of coal, limestone, clay, and hydrocarbons. The Pennsylvanian rocks in this region of Texas have had vital economic importance since the late 1800's; the coal provided fuel, the clay was used in bricks and ceramics, the limestone was crushed for aggregate, and oil and gas were yielded from the subsurface.

Sediments eroded from the young Ouachita, Wichita, and Arbuckle mountains were carried westward and southward across a flat shelf on their way to deposition in the deep basins to the west. Meandering rivers deposited sand in channels, and mud on the adjacent flood plains and deltas, while the lush vegetation on the wet, flat terrain fell and was buried to accumulate coal. As often as the encroaching sediment pushed back the sea, the sea in turn waged battle back to cover the flat land, leaving a record of its passing in the limestone it deposited in defiance of the rivers. The battle between sea and stream waged to and fro for countless cycles. But the streams eventually won the war and the shallow sea was filled, as we have seen, in Permian time. However, it was the back-and-forth cycles of river and ocean that laid down coal, sandstone, mudstone, and limestone together in this area, which later turned into a bonanza of earth-resources.

Pennsylvanian orange/yellow sandstone beds overlie reddish brown shale beds in eroded cuesta country near the Texas 108 crossover. Sandstone and shale cuestas are common between Texas 108 and the

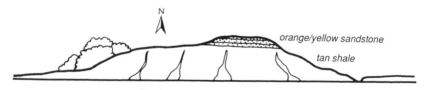

Outcrops of colorful sandstone and shale of early Pennsylvanian-Strawn group, north of Interstate 20 at Texas 108 junction and east of Thurber/Gordon exit.

218

Brazos River crossing. A large quarry, the Gilbert Pit, can be seen south of the highway west of the Brazos River.

The Brazos River forms one of the major drainage systems across Texas. It heads in the high plateau country of the Panhandle and flows across the caprock edge, where its tributaries have cut magnificent canyons. The Brazos then flows across the gypsum plains, red bed terrain and rolling plains of northcentral Texas, before it crosses limestone terrain of the Cross Timbers on its way southeast. The river's last run takes it through the coastal plain before it enters the Gulf of Mexico at Freeport, Texas. Stephen F. Austin believed the Brazos would be the chief waterway of commerce for Texas, and indeed, steamboats and flatboats plied up and down the brown Brazos until railroads took away the trade later in the 19th century. The Brazos River now provides power, water, and recreation via the many dams and lakes along its length.

The edge of the Cretaceous rock cover is just east of the Brazos River, but is difficult to see from the interstate because there is no real topographic edge, nor are outcrops prominent. A careful look at soil color might distinguish the transition: Cretaceous limestones and sandstones are light-colored and weather nearly white, whereas the Pennsylvanian limestones are gray and weather to buff-tan colors. Vegetation cover, limestone rubble in fields, rolling topography, and a few erosional hills with ledges of nearly white limestone, constitute the Cretaceous geologic scene between the Brazos River and Ft. Worth.

U.S. 84
Coleman—Abilene
52 miles

Between Coleman and Abilene, U.S. 84 crosses three bands of Permian rocks, as shown on the northcentral Texas geologic map, and passes by high hills of younger Cretaceous rocks. South of Abilene, the prominent hills on either side of the road are part of the Callahan Divide. This divide is a remnant, erosional plateau of Cretaceous rocks that once covered much more terrain in north-central Texas, until erosion removed this cover from most of the area.

Permian rocks are not very well exposed along the northern half of the road, but the road passes through several good Permian roadside cuts nearer Coleman. At three miles and at sixteen miles northwest

Roadside exposure of Permian Elm Creek formation, of the Wichita group, 3 miles northwest of Coleman.

Close up view of fossils at above location. Black bar is one inch long.

16 miles northwest of Coleman, Permian marine limestone and mudstone beds have abundant fossils.

Fossil "hash" in above section.

of Coleman are two such exposures, where fossils of brachiopods, clams, crinoids, and fossil "hash" are abundant in thin-bedded, gray limestones, which are interbedded with soft shales. The fossils are a nice shallow, clear water, marine assemblage, so they tell us the type of environment in which these limestones and shales were deposited. Although the Permian section tilts gently to the west, the half-degree it does tilt is not discernible at the scale seen in a single roadcut. Thus, the layers in these cuts look quite flat-lying.

Between the two Permian roadcuts, at Silver Valley about nine miles northwest of Coleman, Highway U.S. 84 passes through the edge of a Cretaceous ridge. A water tower is on top of the ridge and you can see where the light-yellow to tan-colored limestone has been quarried at one time.

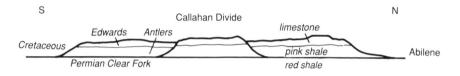

Geology of Callahan Divide, 16 miles south of Abilene on US 84.

Fifteen miles south of Abilene, the road runs through a big gap in the Callahan Divide, where high hills frame both sides of the highway. Yellow-tan lower Cretaceous Edwards limestone forms the resistant cap on top of the ridges. Pink and gray shales and sandstones of the Antlers formation on the lower slopes, also lower Cretaceous in age, rest directly on red Permian rocks at road level. The red soils are indicative of the red Permian bedrock here.

On the west side of the gap is a small historic town, called Buffalo Gap, built on the rich prairies where buffalo once commonly ran through the gap in the Callahan Divide. The pioneer town is now operated as a historic village where visitors are treated to a bit of the old west.

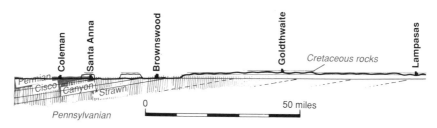

Cross section along US 84 between Coleman and Lampasas. The tilt of
Permian and Pennsylvanian rocks is greatly exaggerated.

U.S. 84-183
Coleman—Brownwood—Lampass
102 miles

Between Coleman and Brownwood, U.S. Highway 84 travels on flat
country underlaid by Permian and Pennsylvanian rocks, then climbs
through hilly terrain of the Lampasas Cut Plain, built of Cretaceous
limestone, on the way to Goldthwaite and Lampasas. The flat area
around Brownwood is in eroded bottomland of the Colorado River
drainage system, hence the reason for the relatively uncommon
exposures of Permian and Pennsylvanian rocks. Look carefully in
stream cuts, roadcuts, and outcrops right around Brownwood to see
tan, yellow, reddish sandstone and limestone beds of Pennsylvanian
age. But the topography, mainly in the form of high mesas standing
above the general level of the countryside, is formed of limestone
layers of younger Cretaceous age that have been etched and eroded for
over ten million years by the Colorado River. As the Edwards Plateau,
south of this area, was elevated along the Balcones fault about ten
million years ago, the ancient Colorado River began downcutting
through the underlying Cretaceous limestones. The carving continued unabated until the Cretaceous layers were breached and eroded
back, exposing the older, tilted Permian and Pennsylvanian rocks.
Isolated mesas of Cretaceous limestone were left standing as lone
sentinels above the river. The mesa by Santa Anna, and the hills west
of Brownwood are examples of these Cretaceous mesas.

The pattern of river drainages shows very well how the streams
have cut headward, carving out the Cretaceous overburden, and
leaving high mesas in the drainage divides. It is also clear that the

Callahan Divide south of Abilene is a drainage divide—rivers on the south side of the divide spill into the Colorado River while water from the north side of the Callahan Divide flows northward to join the Brazos River.

Southeast of Brownwood, the road leaves the Colorado River lowlands and begins a gentle climb into the Lampasas Cut Plains, where hilly, rolling terrain characterizes the roadside scenery. Limestone beds of Cretaceous age are commonly seen on ridges and hill top edges. In a quarry north of Goldthwaite are marly, yellow-tan limestone beds, which are equivalent in age to the dinosaur track beds seen at Dinosaur Valley State Park near Glen Rose.

The limestone is hard and fairly resistant to erosion, so it is a ledge-builder in this area. Note old buildings constructed of limestone blocks in the small towns, particularly in Lampasas. Pioneer folks liked to build sturdily, but they were smart and didn't drag their heavy building stones from long distances, if good stone was locally available. So, you can see a lot of local geology in old house building stones.

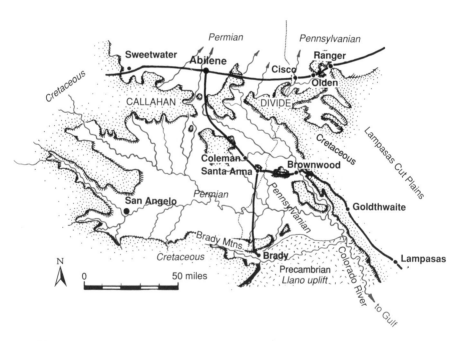

Sketch map shows how Colorado River system has cut through cover of Cretaceous rocks, leaving high-standing Cretaceous mesas, and exposing older Permian, Pennsylvanian, and Precambrian rocks in low river bottoms.

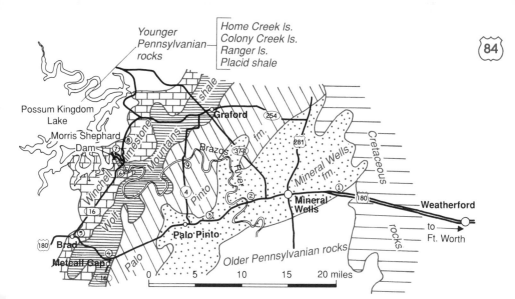

Geologic map of Possum Kingdom area, with road locations. Northeast to southwest bands of Pennsylvanian rocks are covered by Cretaceous rocks east of Mineral Wells.

U.S. 180—Texas 16 Loop
Weatherford—Mineral Wells—
Possum Kingdom Lake

This loop drive across the western Cross Timber country west of Ft. Worth is a geological delight. The countryside itself is a marvelous blend of rolling hills, river bottoms, and steep-cut canyons, where roadcuts, outcrops, and canyon walls display a marvelous variety of rocks.

Mostly what you will see on this drive are sedimentary rocks - sandstones, coals, shales, limestones—of Pennsylvanian age. These rocks were originally deposited about 280 million years ago along a flat shelf area which occupied a notch between the Wichita, Arbuckle, and Ouachita mountains, north and east, and the deep Permian basin, to the far west. Rivers brought sand and mud from the mountains and deposited them across the shelf. Similar to the modern coastal plain of Texas, the Pennsylvanian coastal plain was built of deltas, rivers, floodplains, beaches, lagoons, and barrier bars. In the adjacent shallow marine sea, mud was deposited for some distance across the shelf, brought there by the rivers. In the clearer water

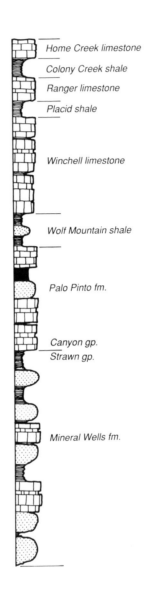

Home Creek limestone

Colony Creek shale

Ranger limestone

Placid shale

Winchell limestone

Wolf Mountain shale

Palo Pinto fm.

Canyon gp.
Strawn gp.

Mineral Wells fm.

Sequence of Pennsylvanian rocks in the Mineral Wells—Possum Kingdom Lake area, Northcentral Texas.

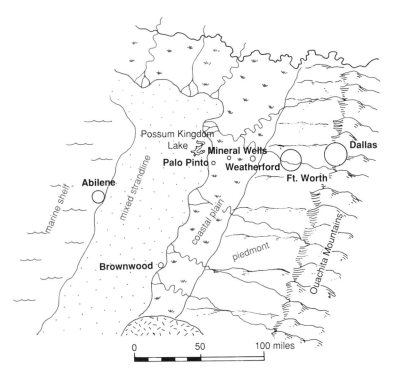

Map of the environments in northcentral Texas during Pennsylvanian period.

farther offshore, marine creatures contributed their calcareous shells to build low limestone banks. Some sediment was even transported beyond the shelf and deposited on the slope and floor of the Permian basin, enough, in fact, to completely fill up the basin by Permian time; it only took a few million years!

Traveling west from Weatherford, through Mineral Wells, Palo Pinto, to Brad, U.S. 180 crosses a series of cuestas held up by hard sandstone and limestone beds that tilt gently toward the west. From Weatherford almost to Mineral Wells is the location of basal Cretaceous Trinity group sandstones, that are important subsurface aquifers. Soils on this sand are also great for growing peaches. Though limestone and sandstone crop out most commonly because they are resistant to erosion, soft mudstone forms the bulk of the sedimentary section. You just don't see the mudstone because it is either covered by vegetation on lower hill slopes or lies unseen in the river bottoms. The rivers are where they are because they preferentially erode into the soft mudstones. Because of the westward tilt, the road encounters

W   E

Cross section arrangement of Nearshore sands, shelf muds, and offshore limestone banks in outcrops in Possum Kingdom Lake area.

a stack of rocks that are older in the east and younger to the west. But also from east to west the traverse first encounters the rocks originally deposited near-shore, on the coastal plain, and in deltas and rivers, while westward the rocks are chiefly limestones and mudstones, originally deposited farther offshore. Study the diagrams to get a good mental picture of the Pennsylvanian geography and environments before heading out on this drive.

Geologists have studied the rocks in the Possum Kingdom Lake area quite intensely for a long time, mainly because the rocks produce important resources. Oil and gas, water, coal, and clay have been gleaned from these rocks for years in this part of the state. Geologists, by studying the rocks in detail at the surface, can then do a better job of predicting where to look for economic deposits in the same rocks in the subsurface. As a consequence, much geologic literature and nomenclature have arisen to describe the Pennsylvanian rocks in Possum Kingdom country. Some readers will be interested in the nomenclature, others not. Some names have been included in this section for those interested, but it's also all right to ignore the names and concentrate on the rocks, enjoying an understanding of their place in the Pennsylvanian world, for example, or relishing the discovery of fossils, or noting how erosion and the rocks have interacted to create the lovely landscape. All of the later observations can be made quite nicely without names, of course. By including some of both approaches, the reader can choose—both are valid ways to do field geology.

U.S. 180 west of Weatherford travels across rolling countryside underlaid by limestone, shale and sandstone of Cretaceous age. This area is part of a Cretaceous rim that stands high and surrounds the lower landscape to the west where the Brazos River has etched its way downward through the Cretaceous to expose older Pennsylvanian and Permian rocks. Limestone rubble in fields and shaly limestone and sandstone in a few roadcuts about ten miles west of Weatherford are the sole indicators of the Cretaceous bedrock.

The environment in northcentral Texas during Pennsylvanian time (280 million years ago.

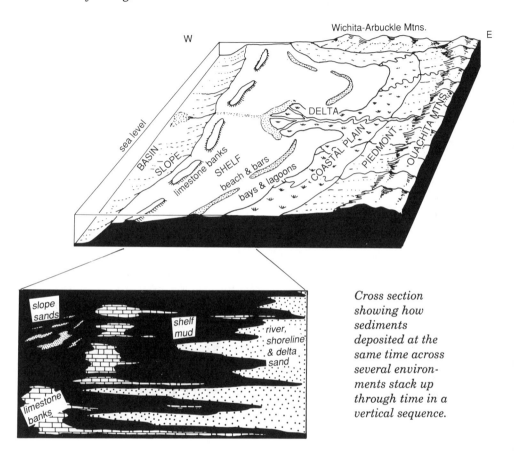

Cross section showing how sediments deposited at the same time across several environments stack up through time in a vertical sequence.

Thirteen miles west of Weatherford the road descends toward Rock Creek. The low hills on the east side of the river represent the edge of the Cretaceous plateau. On the west bank of Rock Creek is a beautiful roadcut (Location 1) with vertical walls, where thick and thin sandstone beds are mixed with thin shale beds. Ripples and cross-beds are nicely displayed in these Pennsylvanian sandstones. The ripples and cross-beds indicate variable water currents moved the sand. Coupling this interpretation with the thin, flat character of the beds, alternating with shale streaks, suggests these rocks were originally deposited in a shallow water environment such as a bay or lagoon, or maybe a tidal flat, as shown on the eastern part of the environmental diagram.

Location 1. Roadcut at Rock Creek. Ripples and crossbeds in Pennsylvanian sandstone (Mineral Wells formation).

Around Mineral Wells, note the cuestas with sandstone ledges north of town. The U.S. Brick plant and clay quarries are prominent north of the highway. Pennsylvanian clay is an important resource in this area, as previously mentioned. Brown sandstone and gray shale cuestas continue west of Mineral Wells.

At the Brazos River crossing, a large roadcut on the east bank (Location 2) is topped by a thick sandstone channel which rests on gray clay. Note how the lower clay slopes are deeply dissected by

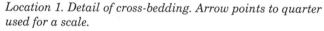

Location 1. Detail of cross-bedding. Arrow points to quarter used for a scale.

Location 2. Exposure of an ancient river channel sandstone (Mineral Wells formation) at Brazos River crossing.

erosion. The sand channel and clays represent deposits of river and floodplain, shown on the eastern half of the Pennsylvanian environment diagram.

A few miles farther west, look for a hill with a round water tower, where there is a roadside cut through thin, light gray 'crinkled' beds of limestone (Location 3). Lots of marine fossils, including brachiopods and crinoids, are to be found here. This limestone probably represents one of the long stringers of shelf limestone that extended eastward when water level was high and sediment supply was low on the Pennsylvanian shelf. Such a stringer is shown on the bottom of the environment cross section. Similar rocks are seen in small roadcuts around Palo Pinto. Cuesta topography, now involving limestone ledges, stands out north of the road.

Location 3. Thin-bedded, fossiliferous, shelf limestone (Palo Pinto formation) at water tower hill east of Palo Pinto.

231

Location 4. Ridge of Wolf Mountains shale at Metcalf Gap.

At Metcalf Gap, prominent ridges of Wolf Mountains shale are topped by brown sandstone ledges, and the cuestas produce pleasant mesa and valley country (Location 4).

Cross section along Texas 16 near Morris Shephard Dam. Map locations are circled.

Limestone beds are exposed in roadcuts on U.S. 180 east of the town of Brad, and north of town on Texas 16 (Location 5). Fossil algae, brachiopods, clams, sponges, snails, bryozoans, and corals are common biotic constituents. These limestone beds (Winchell limestone) were deposited on the margins of a limestone bank, which was a pile of fossil debris located on the outer margin of the Pennsylvanian shelf.

Location 5. Thin beds of Winchell limestone at Brad roadcut.

Location 7. Handsome cliff of Winchell limestone at Morris Shephard Dam.

The banks were thus not true reefs, that is, built by the in-place growth of frame-building marine organisms, as seen in today's oceans. The thick core of the bank is seen at Possum Kingdom Dam to the north.

From Brad, the highway (Texas 16) heads north across mesa and valley terrain, where abundant juniper bushes grow on the dry limestone surface. About seven miles north of Brad, the highway begins its descent toward the Brazos River Canyon. Roadcuts and quarries give an internal view of the Winchell limestone bank, as the road slices downward.

From the Brazos River Bridge (Location 6), vertical canyon walls 150 feet high hug the skyline to the east, while the Morris Shephard Dam holds back Possum Kingdom Lake to the west. The cliff rock seen here is Winchell limestone, which represents the thick core deposits of a limestone bank, where fossils are abundant. Look for algae, clams, snails, brachiopods, and corals. Wolf Mountains shale is in the river bottom.

For a close-up view of the geology of the limestone banks, take the short paved road to Morris Shephard Dam, where a full section of Winchell limestone is freshly exposed (Location 7). Sandstone and mudstone beds (Wolf Mountain shale) at the base of the cliff are the shelf substrate on which the limestone mound grew. As the ocean migrated inland across the shelf, the water got cleaner and deeper. In response, the limestone beds get thicker and 'cleaner' and less muddy, as they built upward; you can see this sequence on the cliff wall.

North from the dam turn-off, Highway 16 climbs the Winchell limestone cliff; at the top is a scenic turnout where you can look out over the Brazos River canyon (Location 8). The canyon makes tight twists and turns, following a looping, river meander pattern. What this pattern suggests is that the river was meandering across the surface of the limestone, then began to downcut into the limestone. Once downcutting began, the meander pattern was fixed, and the ever-deepening canyon preserved the original meander pattern.

To see the youngest group of Pennsylvanian rocks in this area, which is the upper Canyon group composed of Home Creek limestone, Colony Creek shale, Ranger limestone, and Placid shale, one must go westward to Possum Kingdom State Park, on the other side of the lake. These rocks represent cyclic shifts of the Pennsylvanian sea— back and forth from open ocean limestone deposition to shoreline sand and mud deposition.

A few miles north of the scenic turnout, you can elect to follow Texas 16 northwestward toward Graham, or turn eastward on Highway 254 toward Graford. To return to U.S. 180 from Graford, you can either take Texas 377, which traverses limestone hill country, or head south on FM 4, which again crosses the Brazos River. At the river crossing are cuestas of Palo Pinto Limestone, where fossils of corals, algae, and bryozoans are abundant (Location 9). These rocks were deposited on the edge of a limestone bank.

U.S. 283
Brady—Coleman
52 miles

This north-south road between Brady and Coleman crosses Permian rocks in the northern half, Pennsylvanian rocks for most of the southern half, and a hilly segment of Cretaceous rocks for a few miles north of Brady.

From the junction of U.S. 87-283 near Brady, Highway 283 heads straight north across a plateau of Cretaceous rocks that form such a high, steep-sloped ridge above the plains to the north that the ridge is locally known as the Brady Mountains. The highway climbs the Brady Mountain escarpment through a gap carved by Cow Creek, where limestone quarries offer a glimpse into the internal geology of the ridge.

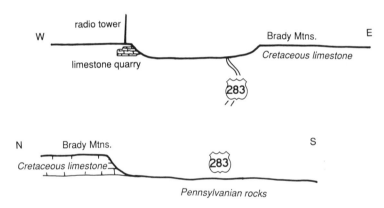

Brady Mountains (2000 feet above sea level) north of Brady on US 283.

After crossing the Brady Mountains, the road travels northward across flat terrain, where occasional rock rubble in fields and blocks of excavated limestone indicate the underlying bedrock is older, Pennsylvanian-age limestone. Thin outcrops of limestone show up on low ridges near the crossing of the Colorado River. But this is flat country—the straightness of the road tells you that—it doesn't have anything to go over or around. The Colorado River here is a mere trickle compared to its expanse of blue-green water southeast of Austin. In side drainages of the Colorado near the small town of Rockwood, low, eroded mesas expose a bit of upper Pennsylvanian sandstone, mudstone, and limestone.

Blocks of yellow-weathered, hard, dense, gray Pennsylvanian-age limestone dredged up from flat terrain, 15 miles north of Brady on US 283.

Between Rockwood and Santa Anna the road rides on slightly younger, lower Permian, limestone and sandstones, though exposures of rock are rare in this flat farmland. Surprising is the high rocky ridge near Santa Anna that looms above the landscape like a lonely ship adrift at sea. Once this ridge of Cretaceous rock was part of a continuous layer connected to the Brady Mountains. All the Cretaceous rock between Santa Anna and the Brady Mountains has

Isolated ridge of Cretaceous rocks at Santa Anna. Edwards limestone is mined to use as aggregate. Lower slopes are composed of lowest Cretaceous sandstone, shale, and conglomerate beds (Antlers sandstone), which directly overlie older Permian rocks at road level.

been removed by erosion and carried away by the Colorado River over the last few million years.

limestone

sandstone, shale

older Permian rocks   Santa Anna

Cretaceous ridge at Santa Anna.

Northwest of Santa Anna, low ridges of gray limestone and reddish mudstone of Permian age are exposed at road level west of Highway 283. From Santa Anna to Coleman, the road continues on Permian rocks, but few are to be seen in the flat terrain.

U.S. 287
Childress—Wichita Falls—Ft. Worth
220 miles

Highway 287, like Interstate 20, runs a transect across the bands of Permian rocks, but misses crossing Pennsylvanian rocks by a few miles, then rides on Cretaceous beds in the eastern one-third as it approaches Ft. Worth from the northwest.

Permian red beds dip gently to the west between Childress and Electra, however, few real outcrops are visible from the highway. Soils are bright red, though, and indicate the presence of the underlying red rocks. Rolling farmland extends in all directions from the highway. Two major bands of upper Permian rocks, the Pease River and Clear Fork groups, are crossed between Childress and Electra.

About 11 miles south of the town of Quanah is Copper Breaks State Park. Colorful beds of the San Angelo formation, part of the Pease River group, are displayed in badlands topography in the park, where the Pease River has carved into the Permian strata on the way to its junction with the Red River. Watch for greenish copper mineralization at Copper Breaks State Park.

A few miles east of Quanah, four distinct mounds are visible, framed against the skyline south of the highway. These are the Medicine Mounds, sacred ground to the Indian people who first

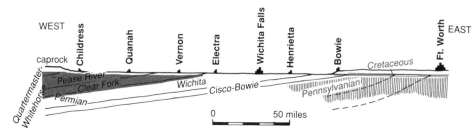

Cross section along US 287 between Ft. Worth - Wichita Falls - Childress. The tilt on the strata is greatly exaggerated.

occupied this area. The mounds are high-standing erosional remnants, and indicate the amount of erosion that has taken place in this area, assuming the regional land surface was once level with the top of the mounds.

Just west of Vernon, Highway 287 crosses the Pease River. This tributary of the Red River carries much red sediment eroded from the surrounding Permian red beds, and contributes significantly to the maintenance of the Red River's namesake image.

Numerous oil fields dot the countryside between Vernon and Wichita Falls. The underlying Pennsylvanian rocks are the source and reservoir for much of the petroleum produced in this area. The Pennsylvanian cycles of coal and limestone deposition produced the rich organic layers to source the oil, whereas alternating layers of sandstone are the reservoirs where the oil accumulated in the subsurface.

At Wichita Falls, the road crosses the Wichita River, another major tributary to the east-flowing Red River. And, not unexpectedly, the Permian band of rocks in this area of the Red River rolling plains is called the Wichita group, though rock exposures are not readily seen from the highway. Most rock outcrops in this region are to be found in steep river embankments, and geologists interested in studying these rocks drive many back roads and hike along streams for miles to get to the 'best' outcrops.

East of Wichita Falls a few good roadcuts can be seen from the highway. A tan sandstone exposure lies on the south side of the road on the west side of the town of Henrietta. About 18 to 23 miles east of Henrietta, oldest Permian beds show up on cuesta hills and in a quarry cut. The rocks exposed are sandstones and shales, originally deposited in stream channels and floodplains which bordered the

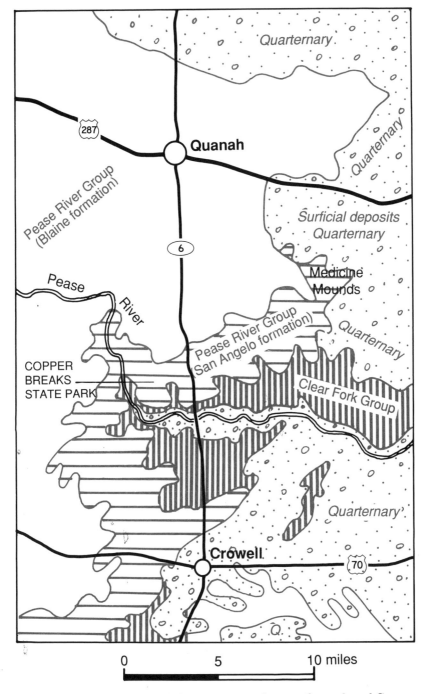

Geologic map of Permian and Quarternary rocks near Quanah and Copper Breaks State Park.

high-standing Ouachita Mountains in Permian time nearly 280 million years ago.

Watch for other roadcut exposures of Permian sandstones west of the town of Bellevue, and on a ridge west of Bowie. The town of Bowie straddles the line between Permian rocks to the west and Cretaceous rocks to the east, though no significant escarpment is seen from the highway to mark the transition.

East of Bowie, at the U.S. 81 crossover, watch for a nice Cretaceous sandstone outcrop on a ridge top above a large concrete retaining wall. The sandstone, more resistant to erosion than adjacent shales, "holds up" the ridge and stands tall above the surrounding landscape. Here this exposed sandstone is the catchment for rainwater which flows downdip into the subsurface where the sand forms an important aquifer in this part of the state. Cretaceous sandstones in this area were deposited in streams and along shorelines, on the edge of a seaway that occupied an area roughly equivalent to the present-day Gulf of Mexico.

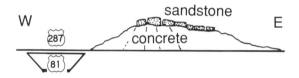

Cretaceous Trinity group sandstone exposure east of Bowie, Texas, at US 287 and US 81 junction.

The road between Bowie and Ft. Worth is on an elevated plateau, called the Western Cross Timbers, which is underlaid by flat-lying, resistant, Cretaceous sandstone and minor limestone beds. Nearer Ft. Worth the topography flattens out in the area known as the Grand Prairie.

Watch for oil pumps and a colorful red, white and pink outcrop of sandstone and shale at the highway crossover in the town of Alvord. At Denton, at the U.S. 287—U.S. 380 junction, is another yellow-tan Cretaceous sandstone exposure on a hill. Southeast of Denton the countryside flattens out on the Grand Prairie, and fertile farmland becomes the main scenery from here to Ft. Worth.

U.S. 377, Texas 144, Farm Road 205
Ft. Worth—Granbury—Glen Rose—
Dinosaur Valley State Park
52 miles

At the I-20 crossover, U.S. 377 heads southwest across terrain underlaid by Cretaceous rocks. A number of good roadcuts and natural exposures are to be seen along this excellent four-lane stretch of road. The highway travels along a plateau of Cretaceous limestone for about five miles, then drops into the Valley of the Clear Fork of the Trinity River at the town of Wheatland. Ridge tops here are held up by nearly flat-lying hard limestones, while the streams rove along on softer marls and claystone beds below the hard limestones.

About fifteen miles down the road, Highway 377 passes through the town of Cresson, located on a ridge crest of hard limestone (Duck Creek limestone) of lower Cretaceous age. Three miles past Cresson are beautiful roadcut exposures, where a set of lower Cretaceous limestone beds can be examined closely. The tan Duck Creek limestone at the top of the cut was originally deposited in shallow marine

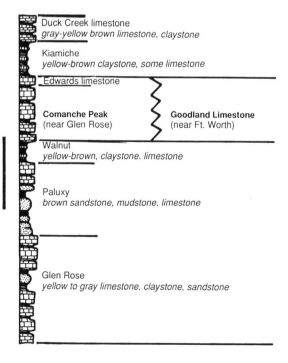

Lower Cretaceous rocks between Dinosaur Valley State Park and Ft. Worth.

Duck Creek limestone
gray-yellow brown limestone, claystone

Kiamiche
yellow-brown claystone, some limestone

Edwards limestone

Comanche Peak
(near Glen Rose)

Goodland Limestone
(near Ft. Worth)

Walnut
yellow-brown, claystone. limestone

Paluxy
brown sandstone, mudstone. limestone

Glen Rose
yellow to gray limestone, claystone, sandstone

100 feet

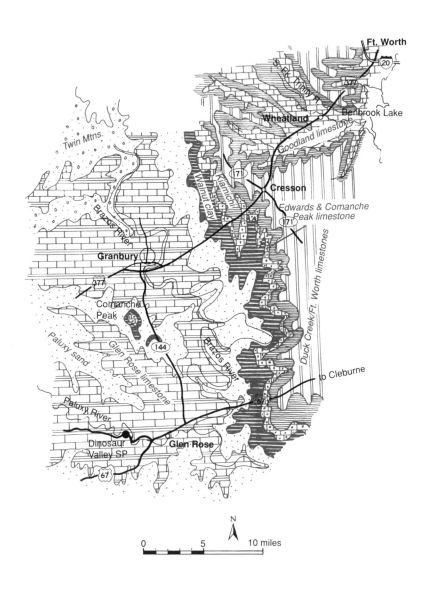

Geologic map of the area between Dinosaur Valley State Park and Ft. Worth.

Roadcut through lower Cretaceous rocks on Highway 377 by Cresson, Texas.

water in early Cretaceous time, 120 million years ago. Oysters and ammonites (large mollusks that look like snails coiled in one plane) are common fossils found in this limestone. The center section of gray, clayey, marly, nodular-looking Kiamichi limestone was laid down in shallow marine water and marshy areas near shore. Oysters and Pecten clams (the Shell Oil clam!) are common fossils. The lowest, hard Comache Peak limestone beds were deposited in clear, shallow marine water in lower Cretaceous time, and contain abundant fossils, including clams, snails, ammonites, oysters, and echinoids. Though Comanche Peak formation forms most of Comanche Peak, Edwards limestone caps the hill which you see as you drive south from Granbury on Highway 144 to Glen Rose.

A large limestone quarry can be seen beyond this roadcut, off to the northwest. After passing the roadcut and quarry, Highway 77 descends into the Brazos River valley. Natural bluffs of limestone extend out from the highway to the northwest. The cliff-forming rocks are the same ones so nicely exposed in the roadcuts and quarry.

Near Granbury, the road crosses a long bridge over the dammed Brazos River. The cliffs on the west side of the reservoir are tan-white limestones of the lower Cretaceous Glen Rose formation. The Glen Rose essentially covers the area from Granbury to the town of Glen Rose, from which the formation was named. It is the unit which preserves dinosaur tracks at Dinosaur Valley State Park, and elsewhere around the state.

After turning south on Texas 144 at Granbury, the high prominent peak seen to the southwest is Comanche Peak. It stands above the average level of terrain, and is an erosional remnant. Think how much rock has been removed by stream erosion! The Comanche Peak limestone on top of Comanche Peak once extended all the way across the countryside to Cresson, where we saw it at the bottom of the Cresson roadcut. So, even more rock used to be on top of Comanche Peak, but has been eroded away.

The few roadcuts and natural exposures along Texas 144 are all Glen Rose limestone. The typical Glen Rose alternations of hard limestone beds and intervening soft marl or mudstone, weathers to a distinctive "stair-step" topography that characterizes Glen Rose out-crops across central and northcentral Texas. Coming to the town of Glen Rose, turn right (west), go through the town and look for Texas 205, which heads north along the Paluxy River toward Dinosaur Valley State Park. A few miles down Texas 205, turn right into the Park entrance.

Dinosaur track locations in Texas.

DINOSAUR VALLEY STATE PARK

It is impossible to avoid an almost mystical, eerie feeling of timelessness as your feet stand inches from the depression in the limy sand where the huge, two-legged, meat-eating monster planted one foot 120 million years ago. The toe marks are exquisitely clear, and no mistaking it, the beast walked on tiptoe on three viciously clawed toes. The most ephemeral of events, a footprint, is remarkably preserved, a mere whisper in time, frozen in place to last 20 times longer than the human race has been on earth. Amazingly, the wet flats where the animal strode preserved not one, but thousands of tracks, scattered at localities across Texas. Fortunately, the best trackways have been preserved for all to see at Dinosaur Valley State Park.

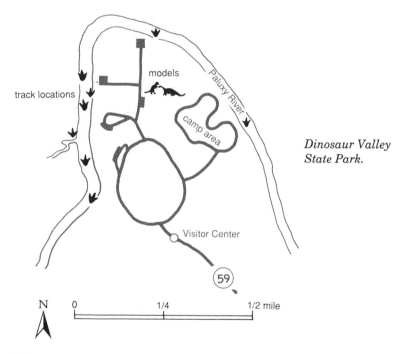

Dinosaur Valley State Park.

What To See

As you enter the park, stop at the visitor center to see the excellent displays before you head out to view the actual tracks. This orientation stop will prepare you to enjoy your track visit even more. The displays illustrate the types of dinosaurs that roamed Texas over 100 million years ago, their biology and family trees, and show how tracks were preserved in the environments of the time. Track casts with the foot bones of the animals that made the tracks is a particularly well done display.

Life-size Theropod *and* Sauropod *models in Dinosaur Valley State Park.*

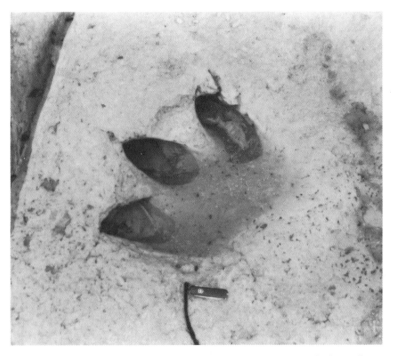

Three-toe track of Theropod *dinosaur. These were made by a "two-legged" carnivore. Knife is 3 inches long.*

The main features of the park are the tracks, of course, along with two full-sized models of a carnivorous *Theropod* and a vegetarian *Sauropod*. These models are located northwest of the camping area, along the north road.

At the end of the north road is an overlook where you can peer down on the Paluxy River bed, to see not only tracks, but several distinct trackways. Sometimes riverbed silt covers the trackways and they are difficult to see. Both washtub-like tracks of *Sauropods* and three-toed tracks of *Theropods* (two-legged carnivore) are preserved here. The trackways march across the riverbed away from the overlook.

Many individual tracks are found on the sandstone layers at river level along the west loop of the Paluxy River. Wander around for awhile to see if you can spot both types of tracks preserved here. Directly across the river from the stairs and northwest parking lot is a set of sharply defined three-toed tracks, recently exposed by the bank-carving action of the river. Note the gray mudstone layer in the

Washtub sized track of "four-legged," Sauropod *dinosaur. Note toe nail marks and rim of mud pushed up around the edge of the footprint. The knife is three inches long.*

bank that once covered the sandy track bed before erosion stripped the mud to expose the track. This little rock sequence graphically shows how the tracks were preserved: the mud was laid down immediately and gently over the track to form a tough, protective layer that lasted for millions of years, until the Paluxy River removed it only a few years ago!

The Dinosaurs' Environment

Alternating sandstone, limestone, and shale layers of the Glen Rose formation are nicely exposed in the cliff face along the tight river bend. These alterations are typical for the Cretaceous sedimentary rocks in this area, and tell us a lot about the environment at the time

Top: Theropod *dinosaur and tracks. Bottom:* Sauropod *dinosaur and tracks.*

248

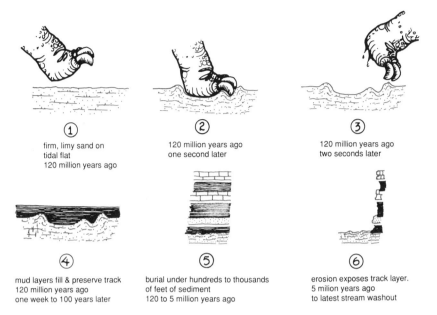

①	②	③
firm, limy sand on tidal flat 120 million years ago	120 million years ago one second later	120 million years ago two seconds later

④	⑤	⑥
mud layers fill & preserve track 120 million years ago one week to 100 years later	burial under hundreds to thousands of feet of sediment 120 to 5 million years ago	erosion exposes track layer. 5 milion years ago to latest stream washout

How tracks are made and preserved.

the dinosaurs walked around here. The Cretaceous sea lay to the south and southeast, roughly where the Gulf of Mexico now is. The Glen Rose area was part of the flat Cretaceous seacoast, which was not unlike today's Texas coastal plain. But, during the Cretaceous the sea was very shallow for a long distance between the seacoast and deeper ocean water. Throughout the Cretaceous, shorelines moved back and forth across this shallow pan as sea level bobbed up and down, driven by the amount of sea water that was being displaced at any one time by gigantic lava eruptions in all the world's mid-ocean ridges. Recall that sea floor spreading was in high gear during the Cretaceous.

During periods of high lava extrusion into the oceans, sea level rose and inundated the continental margins. Shorelines retreated for many miles across the flat pans, marine organisms moved in, and their shells accumulated as limestone. The limestone was deposited directly over the old shoreline sand beds and river silts and muds that once occupied the flat pan. During quiescent episodes of sea floor spreading, little lava emerged, the ocean surface settled back to lower levels, and sea water evacuated the flat pans of the continental margins. Where limestone lay, shoreline sand, and river silt and clay were now spread.

A vivid example of how tracks are preserved.

This picture of back-and-forth shoreline migration is written in the rocky pages exposed in the walls of the Paluxy River. Limestone rests on sand and shale, which in turn rest on another limestone, while above and below, the cyclic rock pattern repeats itself again and again.

The dinosaurs roamed the tidal flats, shorelines, swampy marshes, and lagoons of the Cretaceous seacoast environment. These areas must have been easy trailways, because many dinosaurs walked them

The Cretaceous Glen Rose section along the Paluxy River in Dinosaur Valley State Park.

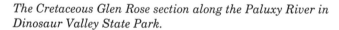

and the paths were firm and flat, as indicated by their tracks—the big beasts didn't sink very deep as they marched along.

Dramas of daily life are told by the trackways. One famous tale is unfurled at the visitor center, wherein a chase involving a dinosaur hunter and its plant-eating prey is written in a long trackway uncovered in the 1920's.

Hunting dinosaur tracks in Dinosaur Valley State Park.

Dinosaur Extinction

Worldwide, the remains of dinosaurs are not found in sedimentary rocks younger than about 65 million years, or about at the level that marks the end of Cretaceous time. Dinosaurs simply became extinct after dominating the earth for 150 million years. Interestingly, dinosaurs were not alone in their grandiose exit, for a host of other animal groups, such as the marine floating plankton, took their final curtain call at this time, too. This seemingly mass extinction of both land and sea-dwelling creatures at the end of Cretaceous time raises the obvious questions of why and how it happened, and scientists have been inordinately creative at answering the questions.

The more popular ideas to explain dinosaur extinction include climate change (it got too cold for them), their food plants died out, diseases finally got them, small mammals ate their eggs, and a meteor impact fouled the air, choking the dinosaurs to death. None of these explanations are entirely satisfying, nor are all of them easy to verify scientifically. How does a scientist study long-departed, non-fossilized diseases of dinosaurs, for example?

On the other hand, good evidence for climate change at the end of the Cretaceous indicates a shift from warm-temperate to colder, seasonal weather conditions worldwide, though the change occurred over a span of a few million years, not instantaneously. The idea of slow, dim-witted, cold-blooded dinosaurs keeling over from frostbite on a late Cretaceous winter eve while the mammals in their furry little coats marched bravely into chilly Cenozoic time, was quite romantic, until paleontologists recently discovered a whole new persona for dinosaurs which includes warm-blood, speedy chases, family life, herd instincts, and nesting accompanied by tender mothering. If dinosaurs were warm-blooded, climate change seems to lose a little of its persuasive power to explain the dinosaurs' extinction.

The idea that dinosaurs left the earth hollow-cheeked from hunger because their food supply ran out is difficult to see in the fossil record. Strange plants did give way to flowering plants during the dinosaurs' reign, but not all at once. Flowers and conifers, for example, were growing for a long time in the Cretaceous and sailed across the dreaded Cretaceous boundary with ease; in fact, the Cretaceous boundary itself is hard to even recognize in the fossil plant record. In other words, the dinosaurs' food base does not look like it changed all that dramatically at the Cretaceous boundary.

Small mammals scuttled around at Cretaceous-end, and from analogy with the eating habits of modern small mammals, some of the Cretaceous brethren were probably egg eaters. The idea that little mammals toppled the mighty dinosaurs by eating all their eggs was kind of cute (and anthropocentric?) when we mammal descendents believed dinosaurs were cold-blooded reptilian monsters. But, if the new evidence for dinosaur nests and mothering has any veracity, would the furry little beggars be able to completely wipe out all dinosaur eggs? Let's face it, mothering means mom keeps nasty, furry things from jumping in the crib. Furthermore, the idea does little to explain why oceanic plankton cashed out with the dinosaurs.

The latest proposal that a large meteor impacted the earth to end the Cretaceous with a loud jolt and a horrifying poof of air-fouling debris has many supporters in the science fraternity. The evidence for the meteor idea is a recently-discovered, widespread layer of iridium-laden sediment at the Cretaceous boundary. Iridium is related to platinum and though a sparse element on earth, it is more abundant in meteors. Barring other mechanisms for iridium concentration, and assuming the iridium is solely meteor-sourced, the implications of the meteor theory are that extinction had to be instantaneous and widespread. It easily links the termination of seemingly disparate organisms such as dinosaurs and plankton, because the blackened air would not only choke the dinosaurs, but kill sun-loving green plankton as well. But wait a minute, what about Cretaceous flowers, conifers, and birds? If the sun's rays were blotted out long enough for plankton to die off in every ocean, why weren't green flowers or conifers decimated? And, if mighty dinosaurs had terminal lung spasms, surely their fragile feathery relatives, the birds, would have fallen off their perches from inhaling the fouled air!

Recent research on iridium-rich layers indicate certain bacteria may be responsible for concentrating iridium and other heavy metals from seawater. Hence, the idea that iridium concentrations indicate a meteor source may not be true at all.

Is instantaneous extinction supported by the fossil record? In Colorado, the three-horned dinosaur, Triceratops, is found well above the last layers of other dinosaur bones in upper Cretaceous shales. And elsewhere, a few dinosaur bones have been found in younger, well-dated Paleocene sediments. Looking backward, we find that the bony-plated Stegosaurus only made it through about one-fourth of the Cretaceous, while the lumbering Sauropods were continually on the decrease throughout the Cretaceous. So, some dinosaurs apparently sneaked over the Cretaceous boundary, while others gave up the race well before the Cretaceous period came to a close. This evidence suggests all the dinosaurs didn't die out all at once, but over a period of time, or at least over longer time than the environmental effects would last from one disaster, such as a meteor impact.

These are fascinating facts and conundrums for geologists to sort through, but the fossil record clearly shows that many organisms, including dinosaurs, died out quite rapidly, considering geologic time, about 65 million years ago. Whether the dinosaurs' demise took a few days or a few years or a few million years, and exactly what the cause, or causes, were to end the long life of this highly successful group of animals is a wonderful mystery story whose ending remains unwritten.

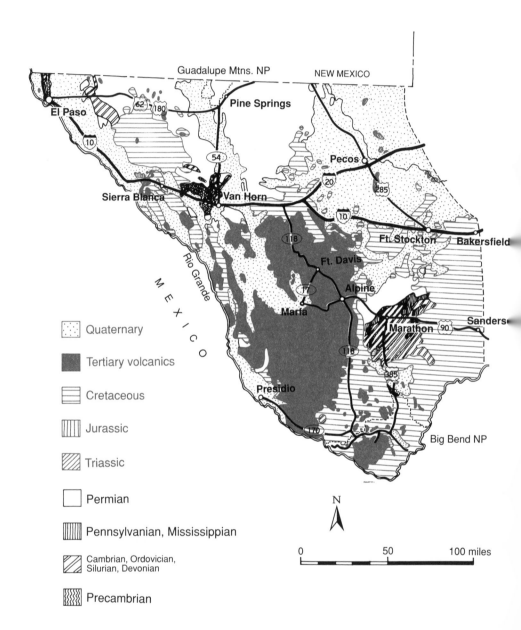

Quaternary

Tertiary volcanics

Cretaceous

Jurassic

Triassic

Permian

Pennsylvanian, Mississippian

Cambrian, Ordovician,
Silurian, Devonian

Precambrian

VI
West Texas
Mountain Country

A Goldilock curl of sand whisks across your windshield as you quietly watch the sun retreat below the black saw blade edge of an unnamed mountain range. The moment is interrupted by a coyote's quick yelp, and something small rustles behind you in the graying shadows of a skeletal yucca. You're in West Texas.

It is a land of sand. It is a land of distance and mountains. It's mood fosters contemplation, and you can't help thinking about rocks, because they're overhead, underfoot, everywhere. If contemplation leads anywhere, it is to the inescapable conclusion that to understand this raw and rough-hewn part of Texas, one must know something of its geology.

Behind the scenery is a vibrant earth-tale of crunching mountains, ripping crustal plates, violent volcanoes. In quieter intervals, miles of reefs grew silently, sediments poured into deep basins, and erosion chewed at uplifted landscapes. Moreover, public tracts have been set aside to preserve much of the story in its natural state—Guadalupe Mountains National Park, Davis Mountains State Park, Balmorhea State Park, and Big Bend National Park.

The oldest rocks in Texas are found in this western corner, in the Franklin and Carrizo mountains where billion-year old Precambrian metamorphic, igneous, and sedimentary rocks appear at the surface. The second oldest suite of rocks, of Paleozoic age (600 - 250 million

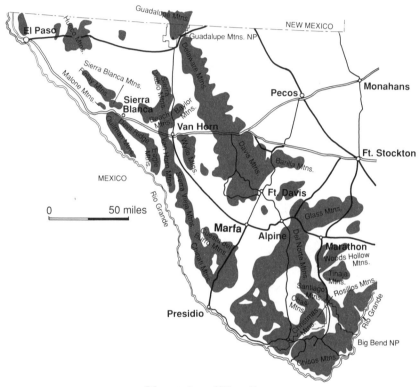

Mountains of West Texas.

years), are grandly laid out north of Big Bend National Park in a tortured string of low mountains around Marathon, Texas, in the so-called Marathon uplift. This uplift is part of the Ouachita Range which sweeps across Texas, mostly in the subsurface passing near the Llano uplift in central Texas, coming to air only at Marathon, and in the Solitario Dome in West Texas. While the Ouachita Range emerged from the collision of two crustal plates about 300 million years ago (when the giant continent Pangaea was assembled), the adjacent crust buckled downward to form deep depressions—the Delaware and Midland basins—in which thick sections of sediment accumulated and around which magnificent reefs grew in profusion. These Permian rocks are seen in the Delaware, Apache, and Guadalupe mountains.

At the end of the Mesozoic period, 120 - 60 million years ago, the supercontinent Pangaea wrenched apart, North America separated from Eurasia; the Rocky Mountains punched upward along the western edge of the continent as the North American plate overrode the Pacific plate. In West Texas, virtually all of the northwest-southeast oriented ranges are the result of this Rocky Mountain building episode.

In the Big Bend of the Rio Grande, the older Ouachita Range meets the trend of the younger Rocky Mountains to create a fascinating, but complex landscape.

About 35 million years ago, western North America stretched somewhat in a relaxation period following the intense compression that built the Rocky Mountains. The crust cracked and faulted, creating down-thrown basins amidst upthrown adjacent mountain blocks, while hot magma rose through the intervening fractures from the mantle below. Volcanoes erupted, spreading lava and ash over a wide area of West Texas. Some bubbles of lava never made it to the surface, but stalled as blisters within the sedimentary rock cover, only later to be exposed by deep erosion. The eruptive volcanic episodes are recorded in the Davis Mountains and Paisano volcano system west of Alpine. Round knobs and hills of volcanic intrusions add their peculiar forms to the scenery of Big Bend.

From the time of volcanic activity in West Texas, 35 million years ago, to today, erosion has been the dominant shaper of the landscape. Look up at Santiago Peak, a volcanic intrusion, to realize how much rock has been weathered, eroded, and removed from this country. Alluvial fans and talus piles at the edge of every West Texas mountain slope also give some notion, even in today's dry climate, of the power of erosion. Quiet, relentless, daily erosion, operating over millions of years, is perhaps the most dynamic geologic process after all! But, it is the interplay of all these geologic forces, each acting in their own time frame, that creates the unique landscape of West Texas.

Caliche

In the semi-arid regions of Texas, mainly in the west and northwest, a white, crusty to nodular, limy soil zone, known as caliche (pronounced ka-LEE-chee), is quite common, notably forming the caprock on many mesas and buttes. Caliche is rich in calcium carbonate, which is precipitated in the soil pores by capillary action from lime-charged ground water.

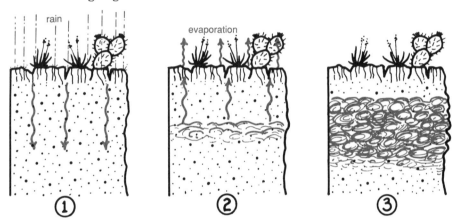

The process works this way: 1. rainwater dissolves lime (calcium carbonate) as it percolates downward through the soil; 2. after the rain, water at the surface evaporates to the dry sunbaked air, causing soil water below to also rise to the surface and evaporate; the lime is left behind as a precipitate; 3. the perennial repetition of this process builds an impervious, cemented zone of crusts and irregular nodules up to several feet thick.

Interstate 10
Bakersfield—Ft. Stockton—Van Horn—
El Paso—New Mexico Border
295 miles

This long segment of Interstate 10 traverses a splendid variety of geology, from limestone mesa country near Bakersfield, across wind-swept flats by Ft. Stockton, past volcanic mountains. It continues through northwest–southeast-trending mountains of the Rocky Mountain system where rocks from old Precambrian to Permian and Cretaceous are wonderfully exposed in the desert climate, then onward through wind-blown sand dunes that parallel the verdant Rio Grande. Finally it goes past block-faulted Franklin Mountains near El Paso, at which point the road heads north toward the New Mexico border.

Higher rainfall near San Antonio creates "softer" mesas on limestone sections.

Whereas "sharp" mesas occur in lower rainfall areas near Bakersfield and Ft. Stockton.

The rocks are the same. The weathering and erosion are different!

Between Bakersfield and Ft. Stockton, the interstate nestles amongst mesas, hills, and buttes as it follows the Old Spanish Trail. This is real West Texas desert country where about 14 inches of rain fall per year. If you have driven from the east on I-10, you will notice the change in shape of the mesas, from a hilly, rounded form common near San Antonio where the rainfall is 30 inches per year, to a sharp-edged, true desert mesa in this vicinity. The rocks are virtually the same; the weathering and erosion are different! In moist regions, chemical weathering causes exposed rocks to disintegrate to pebbles and soil, which creep downslope, rounding off slopes and making "soft" looking hills. With little continuous moisture in the desert, chemical weathering is minimal, so soil is thin. Rocks simply break from cliff edges and the main agent of erosion is high-intensity run-off from water supplied by thunderstorms. "Sharp-edged" mesas of exposed rock are thus a common landform seen in the desert.

Near the junction of U.S. 67 with I-10, about 20 miles west of Bakersfield, the mesas near the highway have limestone caprocks of

Early Cretaceous age. What you are really seeing here, as the mesas get farther apart, and are finally gone by Ft. Stockton, is the western, eroded edge of the Edwards Plateau that stretches back eastward into the Texas Hill Country.

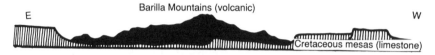

Mountain profile to the south from Interstate 10 a few miles west of Ft. Stockton.

Around Ft. Stockton the road sweeps across flat, gravelly open land of the Stockton Plateau, where alluvial fans and streams have filled up low areas with rubbly debris eroded and transported from surrounding mesas and mountains. West of Ft. Stockton, more mesas with limestone ledges are seen off in the distance to the south. The road cuts through Lower Cretaceous limestone beds that are dipping at quite an angle to the west - evidence of faulting and movement of the rock section here. A few miles farther west, the same rocks are dipping in the opposite direction, toward the east. You have just driven across a syncline!

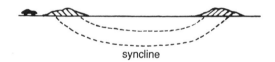

syncline

The Barilla Mountains form a distinctive profile on the southwest skyline. This is a volcanic range, composed mainly of extrusive lava flows resting on gently tilted Cretaceous sedimentary rocks. The Barilla Mountains heaved up about 35 million years ago as part of the Davis Mountain volcanic complex.

More mesas of lower Cretaceous limestone of the Washita group, are seen 25 to 30 miles west of Ft. Stockton, and where the high peaks of the Davis Mountains loom on the southern horizon. Flat alluvial fan deposits spread out northward from the highway. The panorama of the volcanic Davis Mountains parallels the highway for many miles, well past Van Horn. The extent of the panorama is a clue to the vastness of the volcanic field that formed around 35 million years ago by the eruption of a string of volcanoes here in West Texas.

Profile of Davis Mountains from junction of I-10 and I-20 west of Ft. Stockton.

Near the U.S. 290—Texas 17 exit to Balmorhea, Ft. Davis and Ft. Davis State Park, black layers of volcanic rocks are seen next to the highway. The hills are rolling, rubbly with dark rocks. They are "unorganized"-looking, typical of landforms underlaid by volcanic rocks.

Balmorhea State Park, only a few miles from I-10, is a pleasant "oasis" stop. The springs in the park have drawn campers for thousands of years, as indicated by Indian artifacts in the area. As early as 1851, canals were built from the springs to irrigate crops, and in 1935 the Civilian Conservation Corps began to construct the park's swimming pool, concession buildings, and park residence. The

Cross section of intake and recharge area and springs at Balmorhea State Park.

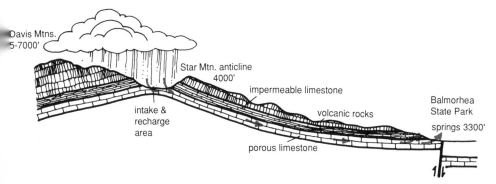

Balmorhea recreation area lies on gravel beds and alluvial fans built outward from the Davis Mountains. Upper and lower Cretaceous limestone, sandstone, and shale deposits form the bedrock around the park.

The springs are charged by rainfall in the Davis Mountains. The water flows downslope, making its way through the cracks and holes and small caverns etched in the limestone beds. It emerges at the surface by spewing up along fractures and faults in the limestone, nearly 700 feet lower than where it entered the aquifer at Star Mountain.

Where I-10 crosses dry rivers, note the mix of limestone fragments and dark volcanic cobbles, just what you would expect to be eroded from the volcanic Davis Mountains and surrounding skirts of Cretaceous limestone beds seen in this area.

West of the intersection of I-10 and I-20, the freeway bisects north-south trending mountain ranges. A long flat plateau on the horizon north of the road west of Kent is the southern flank of the Apache Mountains. They are part of a string of mountains that extend northward to the New Mexico border and include the Apache, Delaware, and Guadalupe Mountains. These ranges expose huge reefs and deeper water deposits of sandstone, limestone, and shale, deposited during Permian time in the Delaware basin. The best place to see these rocks is in Guadalupe Mountains National Park.

Low mesas near the highway between Kent and Plateau are eroded remnants of flat-lying beds of Cretaceous limestone that once covered this area. The roadcut at Plateau is a cross section of the conglomerate and sandstone material that fills the Salt Flat valley.

Between Plateau and Van Horn, I-10 crosses a topographically low desert area, called Salt Flats, complete with blowing sand dunes and salt deposits left over from shallow lakes that existed in wetter times during the Ice Age. The Delaware-Apache Mountains are the uplifted, eastern border of the Salt Flats, whereas the Baylor, Beach, and Sierra Diablo mountains form the steep west flank.

To the south are the Wylie Mountains, another uplifted block of Permian carbonate rocks, which rise sentinel-like from the floor of the Salt Flats.

Three Mile Mountain profile west of Van Horn.

Precambrian rocks in Interstate 10 roadcut west of Van Horn. Dark rocks are metamorphosed sediments (phyllite) and light rocks (pink in roadcut) are metamorphosed rhyolite.

The town of Van Horn is the portal to a beautiful pass between the Carrizo Mountains south of the highway and Three Mile Mountain to the north. Beach Mountain and the Baylor and Sierra Diablo mountains are seen in the distance north of Van Horn.

Cambrian sedimentary rocks, laid down at the dawn of Paleozoic time almost 600 million years ago, form the lower slopes of distinctive Three Mile Mountain, whereas hard Permian limestones form the resistant cap.

Watch for the rest area a few miles west of Van Horn where you can see Precambrian metamorphic rocks next to the picnic tables. Old Precambrian rocks are so rarely seen at the surface in Texas (Llano uplift, Franklin Mountains, Carrizo Mountains), that these outcrops in the Carrizo Mountains take on quite a unique status! Notice the green-black color and dense packing of small, parallel grains in these very hard rocks. They once were mudstones that became heated and pressured at deep burial depths, to be "metamorphosed" into new rock, now called phyllite. White silica veins shoot through the meta-morphic rock in many places.

A highway roadcut just west of the rest stop shows these dark phyllites in sharp contact with pink metamorphosed rhyolite, origi-nally volcanic rock.

Between the Carrizo Mountain pass and the town of Sierra Blanca, the highway rides across flatter desert country, built on the surface of alluvial fans that extend outward from the surrounding mountains. Ranges on the distant skyline to the south are the Quitman Mountains, a Rocky Mountain range uplifted in Late Cretaceous time, about 60 million years ago.

West of Sierra Blanca, rubbly, dark brown hills and rounded mountain topography indicate volcanic rocks again. Three distinctive, conical volcanic peaks north of the highway are Round Top, Little Blanca Mountain, and Sierra Blanca. The road makes a sweeping turn to the north around a large volcanic intrusive. It then descends southward into a broad valley where colorful, pink-white-tan, varicolored badlands are etched into the original smoothly sloping top of the alluvial fans. Here the drainages erode deeply as they approach their confluence with the mighty Rio Grande, now only a few miles away to the south.

Sierra Blanca, Little Blanca, and Roundtop.

As the Interstate swings to the northwest again, it closely follows the course of the Rio Grande all the way to El Paso. Serrated Laramide ranges on the Mexico side of the river are outlined against the sky, while the green fields on the river floodplain nestle at their base. The road runs on a high terrace level above the river; much of the surrounding landscape is windblown sand and dune deposits. Abundant sand in the dry flat of the river bottom is readily available for the wind to distribute. In some places the dunes are stabilized by the sparse vegetation, but look for rounded sand dunes, and wind ripple marks on free sand faces. They are especially obvious in low sun angles of early morning or late afternoon.

Cross section of Franklin Mountains near El Paso.

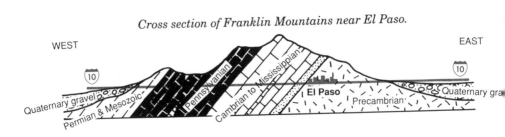

Near El Paso, low hills far in the distance to the northeast are the Hueco Mountains; the Franklin Mountains hover over El Paso to the north. The long stretch of flat desert and dry sand the road has just passed over is the southern end of the Hueco bolson, a low, down-dropped structural area between the Franklin and Hueco mountains. Sedimentary fill in the Hueco bolson is nearly 9,000 feet thick! Hueco in Spanish means hollow or hole; bolson is a geologic term used in the west to describe a structural depression, or basin, having no drainage outlet.

From El Paso to the New Mexico border, Interstate-10 passes the Franklin Mountains, a block-faulted range; a thick section of Paleozoic rocks is stacked up on the west side, overlying a section of Precambrian rocks on the steep east flank. If you look carefully from El Paso, you can see the westward dip of the rocks in the Franklin Mountains.

The highway transects sand and gravel exposed along the roadway as it heads north toward the New Mexico border. These are alluvial fan deposits, which extend toward the Rio Grande from their heads in the Franklin Mountains.

Franklin Mountains looking north from El Paso. Precambrian rocks at far right are overlaid by westward-dipping Paleozoic sedimentary rocks, seen at left.

265

Interstate 20
Monahans—Pecos—Interstate 10
92 miles

Driving I-20 between Monahans and I-10 is a true desert experience. Between Monahans and Pecos is vast, flat country punctuated by sand dunes. Between Pecos and the I-10 junction, there is less blowing sand, but the countryside is still wide, flat, and lonesome.

Interstate 20 crosses the wide, dry Pecos River bottom east of the town of Pecos. On the east bank are low outcrops of varicolored rocks of Triassic age that are not completely covered by the sand dunes. The sand in the Pecos River valley, exposed and dry most of the year, is the source for the sand dunes east of the river. Look for white caliche beds in the area around the Triassic outcrops east of Pecos.

Five miles west of Monahans, the pumps of the North Ward and South Sealey oil fields spread far out from the highway both north and south of the road.

From Pecos to I-10, the road traverses flat sand and gravel deposits laid down during the Ice Ages by streams flowing out of New Mexico highlands to the north.

About four miles east of the I-20/I-10 junction are low mesas of lower Cretaceous rocks on either side of the road. These rocks are part of a ridge uplifted along northwest–southeast-trending faults.

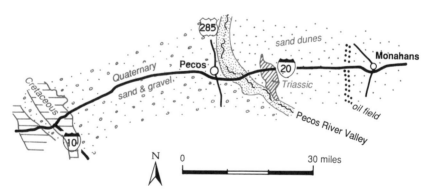

Sketch map of geology along Interstate 20, Monahans to Pecos to Interstate 10.

U.S. 62–180
El Paso—Guadalupe Mountains
National Park—New Mexico Border
130 miles

As the road heads eastward from El Paso, you get a good view to the north of the Franklin Mountains, which rise over 7,000 feet above sea level; they bring old Precambrian granite and metamorphic rocks to their highest structural elevation in the state. More than 10,000 feet of sedimentary, igneous, and metamorphic rocks are exposed in the Franklin Mountains. This north-south range is a tilted, faulted horst block mountain, the steeper side facing east.

The road heads out across desert country for about twenty miles before crossing the Hueco Mountains seen on the skyline ahead. The depression between the Franklin and Hueco mountains is called the "Hueco Bolson." Hueco in Spanish means hole or hollow, while bolson, meaning pocket or purse in Spanish, is a geologic term coined in the west and southwest for a structural depression or basin having no drainage outlet. Usually bolsons are underlaid by great thicknesses of sediment fill, derived from erosion of the adjacent mountains. The Hueco bolson is no exception—9,000 feet of clay, silt, sand, and gypsum are between your tires and solid rock below!

Schematic cross section.

Both the Franklin and Hueco mountains are part of the Rocky Mountain trend, shoved upward as part of the great Laramide mountain building period in late Cretaceous time, about 60 to 70 million years ago. The Hueco bolson collapsed further during the great extensional, Basin and Range structural period that occurred throughout the west; it began in Miocene time, from about twenty million years ago to, in many areas, the present.

The surface of the Hueco bolson is fairly flat, covered by alluvial fan and gravelly stream deposits along with sand hills and dunes. Some sand can be seen trapped in piles behind the sparse sage and creosote

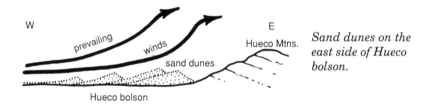

Geologic map along US 62-180 between El Paso and Guadalupe Mountains National Park.

bushes, but most of the sand moves west to east across the bolson where it is piled on the west flank of the Hueco Mountains. The prevailing westerly winds hug the ground and move sand eastward, but when the Hueco Mountains are encountered, the winds rise, lose their velocity, and drop sand. Yes, geology even influences the wind!

Sand dunes on the east side of Hueco bolson.

You see panoramic views ahead of the Hueco Mountains, where Cerro Alto is the highest peak, 6,717 feet above sea level. The bolson surface is about 4,000 feet in elevation. As the road approaches the foothills of the Hueco Mountains, the hills you see north of the highway are composed of Hueco limestone of Permian age. You'll see a lot of the Hueco limestone from here to the east side of the mountains. Good roadcuts of steeply dipping westward-tilted Paleozoic limestones and shales are seen on the western flank of the Hueco Mountains. These are Pennsylvanian Magdalena limestone and underlying Mississippian Helms formation.

Both these units are seen clearly on Helms Peak on the south side of the road. The Magdalena is particularly interesting for its marine to nonmarine origin, wherein petrified wood, algae mounds, chert nodules, and crinoid fossils are all found. A quarry near the Hueco Tanks road (exits north) is in Magdalena limestone, mined for its very pure calcium carbonate.

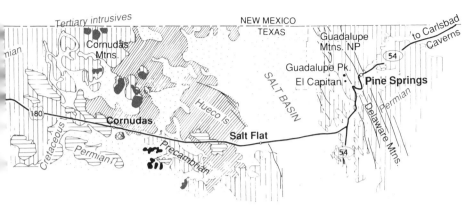

The roadcuts get bigger and the surrounding hills taller as the road passes through Pow Wow Canyon. The Magdalena limestone is predominant in these cuts, but as you continue eastward, Permian Hueco limestone exposures prevail. Notice at the roadside park in the canyon that the limestone beds now tilt toward the east. Within the lower Permian section is an angular unconformity, where flat-lying beds lie directly over tilted strata.

Out of the canyon, but still on the east flank of the Hueco Mountains, Hueco limestone makes up the undulating topography, created by the irregular solution of the underlying limestone by the relentless dissolving action of rain water.

About 45 miles east of El Paso watch for white Cretaceous limestone mesas on either side of the road. These marine limestones once covered the countryside, but have been uplifted, and severely eroded, leaving only these remnant mesas as hints of their former extent. Ten miles east of Cornudas the road cuts another small patch of these rocks.

At Cornudas the rounded hills to the north are the Sierra Tinaja Pinta, which are 35-million-year-old igneous intrusions—small outliers of the same huge volcanic event that created the Davis Mountains near Big Bend National Park. On the far northern horizon, round hills on the Texas-New Mexico border are the Cornudas Mountains, also cored by Tertiary intrusive bodies.

The road passes low outcrops of Permian-age limestone, shale, and siltstone east of Cornudas, and then treks out across the white, white salt flats. The Salt Flats are the dry salt remnants of lakes that once shimmered in the bolson, fed by runoff and springs from the Guadalupe and Delaware mountains during wetter times in the Ice Age about one million years ago. Lest you think these are worthless salt pans of no consequence, a war was fought out here in 1877 over the rights to

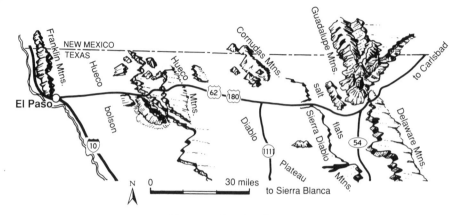

Topography sketch map.

mine, transport, and market this salt! The term "salt-of-the-earth," indicating a person's basic value, hence takes on special meaning after you have driven the Salt Flats of West Texas.

Ahead is the magnificent west cliff-face of the Guadalupe Mountains, and stretching to the south, the Delaware Mountains. The rise of the Sierra Diablo can also be seen stretching to the south, forming the west flank of the Salt Flat. The Salt Flat is a downdropped bolson between the Sierra Diablo and Guadalupe/Delaware mountains.

A few miles east of the junction of U.S. 62-180 and Texas 54, the road climbs Guadalupe Pass into Guadalupe National Park, where the geology of a giant Permian reef is magnificent, unique, and well exposed in roadcuts, cliff faces, and outcrops. (See section on Guadalupe National Park.)

From the Park to the New Mexico border, Highway 62-180 follows the high scarp created by the Permian reef rocks, seen north of the road. Roadcuts for several miles east of the park are cut from flat-bedded rocks originally deposited in deep Permian sea water at the front of the reef. Spectacular banded anhydrite and calcite beds of the Jurassic Castile formation are seen at the state line.

U.S. Highway 62-180 continues into New Mexico and Carlsbad Caverns, which are located at the east end of the Guadalupe escarpment. The reef rocks were lifted out of their deep burial place, and in the process, groundwater etched its way into fractures in the limestone to create these marvelous caverns. From El Capitan on the west to Carlsbad Caverns on the east, the geologic story told by the rocks on this long escarpment is a fine one indeed.

Quarternary conglomerate on the east side of Marfa.

U.S. 90
Marfa—Alpine
26 miles

Have you ever driven through a volcano? No? Then get ready, because you are about to do just that as you travel U.S. 90 between Marfa and Alpine.

The west half of the road is across the flat, alluvial plain of the Marfa Basin, whereas the eastern half of the road cuts through the Paisano Plateau formed of rocks from the Paisano Volcano that erupted in Oligocene time about 35 million years ago.

Heading east from Marfa, the road crosses Alamito Creek, at the east end of town, where an outcrop of alluvial fan conglomerates and stream gravels is nicely exposed south of the highway. Note all the different kinds of rounded cobbles and pebbles that were deposited here as debris from the surrounding mountains.

Puertacitas Mountains form the skyline to the north and the volcanic mountains of the Paisano Plateau are on the skyline to the east. The distinctive profile of Cathedral Mountain (6,800 feet) is very clear to the southeast.

Cathedral Mountain (6,800 ft. elevation), an intrusive mass, as seen looking east from Marfa Basin.

Thirteen miles east of Marfa a historical marker explains the significance of Paisano Pass. The Spanish explorer, Juan Dominguez de Mendoza, camped here on January 3, 1684. After 1850 this spot was well known on the Chihuahua Trail, an emmigrant road to California.

From Paisano Pass eastward to Alpine, U.S. 90 passes marvelous roadcut exposures that show the internal workings of a volcano that erupted about 35 million years ago. If the magma feeding a volcano has a lot of water or gas in it, the eruption will generally blast forth explosively, spreading ash and various sizes of particles and debris over a wide area. If, however, the magma contains very little gas or water, it will ooze to the surface and flow laterally from the vent as a hot liquid, or lava flow. Lava may also move upward along fractures and cracks to produce walls of hardened rock called dikes. And sometimes, the lava never makes it to the surface but injects a plug that penetrates into some rock but does not completely cut through the overlying section. It is not uncommon for the caldera of a volcano to collapse after magma has been removed from below; it forms a broken jumble of angular blocks, breccia, that collect in the volcanic neck. All these aspects of a volcano are recorded in the U.S. 90 roadcuts west of Alpine.

The roadcuts along U.S. 90 west of Paisano Peak are mainly in the collapsed caldera of the Paisano Volcano. Much broken and fragmented rock is seen along here. At the roadcut where the Baptist encampment road enters Highway 90 is a large dike that cuts broken

Aspects of volcanic deposition.

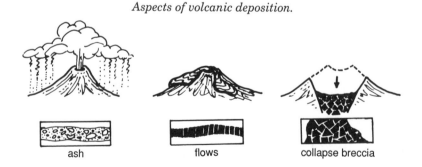

ash   flows   collapse breccia

272

rocks of the collapsed zone. This three-mile-long dike is the largest in the Paisano Volcano.

Just east of the collapsed caldera is a zone of lava composed of very light-colored rhyolite, which is a very fine-grained version of granite. Rhyolite is fine-grained because it came from molten rock that was extruded to the surface where cooling was so rapid no big mineral crystals had time to form. Granite, on the other hand, came from the same type of molten rock soup as rhyolite, but having solidified below the surface, it cooled much more slowly, and consequently grew large mineral crystals. If you look at a piece of granite, you'll see mostly the minerals pale feldspar and gray quartz, and lesser amounts of dark mica or hornblende. In a microscope you would see the same suite of minerals in a piece of rhyolite.

Geologic map of Paisano Volcano west of Alpine. Paisano Peak is at location P. Circled numbers are keyed to photographs of roadcuts at these locations.

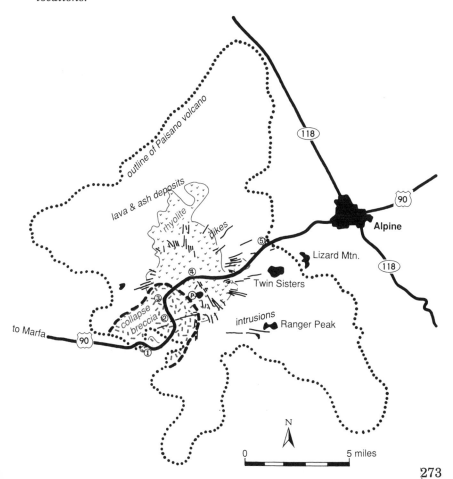

1 on the map. Dark dike cuts light rhyolite in collapsed terrain.

2 on the map. Baptist encampment dike, 3 miles long, is largest in Paisano Volcano. The dike cuts collapse breccia.

3 on the map. Classic collapse breccia. Bar is one foot long.

*4 on the map. Dike cuts rhyolite. Arrow points to edge of dike. Note stria-
tions on dike face which record flow motion along the edge of the molten
rock as it intruded the rhyolite.*

*5 on the map. Dark volcanic flows at east edge of Paisano Volcano. Vertical
joint columns are distinctive.*

View west from Alpine to Paisano volcanic field on skyline. Twin Sisters intrusion left of center. Ranger Peak, another instrusive body, to far left.

The remainder of the Paisano Volcano, as shown on the map, is made up of very dark-colored lava flows and ash deposits.

Looking to the west from Alpine you can see the profile of the Paisano Volcano on the skyline. The museum on the campus of Sul Ross State University in Alpine is open to the public and has geology displays as well as history and other science exhibits.

U.S. 90
Sanderson—Marathon—Alpine
85 miles

The eastern two-thirds of U.S. 90 between Sanderson and Marathon winds its way through canyons and mesas built of flat-lying sedimentary rocks of early Cretaceous age. The remaining third of the roadway around Marathon crosses folded, faulted, and uplifted Paleozoic rocks that are part of the old Ouachita Mountain chain.

West of Marathon, thick cliff sections of tilted Permian rocks form low block-ranges on either side of the highway. Closer to Alpine, Cretaceous rocks are again encountered, but Alpine itself rests on volcanic rocks of much younger age than the rest of the rocks seen along the entire stretch between Sanderson and Alpine.

Cross section from Sanderson to Alpine on US 90.

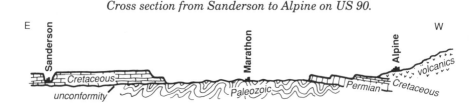

Cretaceous section at Sanderson.

Santa Elena-big gray limestone cliffs dominate hills

Sue Peaks-slope, soft limestone and shale
Telephone Canyon-firm limestone at river level

At the west edge of Sanderson is a large quarry where rock was removed to build the earthen dam seen north of U.S. 90. The tan color of the fresh limestone contrasts with the weathered, gray, natural outcrops of the same rock. The rock units, or formations, seen here in Sanderson Canyon and in the quarry are the same ones seen far to the southwest at Santa Elena Canyon west of Big Bend National Park. The conditions for the sedimentation of these rocks were spread over a wide area.

Roadcuts a few miles west of Sanderson, where the canyon widens to a broad valley, give a close-up view of these Cretaceous limestones.

In the wide valley west of Sanderson, alluvial fans spread outward from mesas. Roadcuts are cross sections through these fans, and display the sand and gravels deposited to form the fans. This area is "classic" desert topography, and shows how desert erosion interacts with the bedrock geology to create the unique mesa valley landscape. Normally dry streams carry immense amounts of water in short bursts after rainy thunderstorms pass. It is then that virtually all the

Cretaceous sedimentary rocks in the Sanderson quarry.

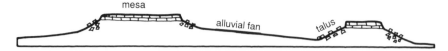

Typical topography west of Sanderson.

erosive work is done in the desert. It seems like a lot is written about "wind erosion" on public signs and displays, but the wind mostly polishes rocks and builds sand dunes, certainly significant in the desert. But, it is water in intense bursts that really carves the desert landscape and creates the landforms.

Talus slopes below flat-lying hard "caps" on mesas are typical in the desert. Rocks forming the caps, whether limestone, sandstone, or even lava, are usually fractured and "jointed"; blocks periodically fall off the mesas edge to form talus slopes below.

About 35 miles west of Sanderson watch for the first blasted roadcut, because here the geology really changes. To the south is House Mountain, a large flat mesa of Cretaceous limestone. In the roadcut, however, are dark, gray, hard limestone beds that stand up at a high angle! You are seeing the exposed edge of the old Ouachita "mountains," where the Dimple limestones of early Pennsylvanian age (315 million years) stands upright in a northeast-southwest trending fold. This locality also represents a major "unconformity" where old rocks were eroded and much younger rocks were deposited over them.

Less than a mile down the road begins a series of beautiful roadcuts where Paleozoic rocks stand literally on end. These vertical beds of alternating sandstone and shale, the Tesnus and Haymond formations, first tell a structural geology story of intense deformation in a

Dimple limestone on US 90.

Close-up of Dimple limestone on US 90.

Paleozoic mountain-building episode. They also tell a fascinating tale of deposition prior to their uplift.

The sand beds were deposited in the deep sea by turbidity currents, mud-sand-water slurries that flow down continental slopes and canyons at high speeds and deposit their load on the deep sea floor. Some modern turbidity currents have been clocked at over 50 miles per hour! Earthquakes probably trigger most turbidity currents by jarring loose the soft sediment lying on the outer continental shelf and upper continental slope. This sediment then flows downslope as a turbulent mass, driven by the force of gravity.

Tesnus formation.

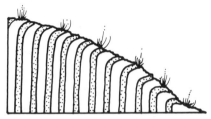

Vertical sandstone beds (Tesnus formation) show curved effect of downslope soil creep.

Each sand bed you see in the roadcut represents the deposit from a single turbidity current. The shale in between is the product of gentler settling of the very fine mud stirred up by the turbidity current, as well as the fine "rain" of mud continuously dumped out to sea by the outflow of river waters. An amazing 14,000 feet of sediments stacked up on the sea floor in this area during Paleozoic time!

Originally, of course, all these sand and shale beds were laid down horizontally. So the fact they are now standing upright means they must have been subjected to epic mountain-building forces. And, indeed they have. The uplift and faulting and folding of these rocks took place about 300 million years ago in late Pennsylvanian to early Permian time, as part of the great Ouachita mountain-building episode. Ridges of the folded remains of these mountains can be seen running across the countryside to the south of Highway 90.

Near the town of Marathon, an east-west ridge north of the highway shows how abruptly the direction can change in the mountain folds. Capping the ridge is the white, silica-hard Caballos novaculite.

U.S. 385 heads south to Big Bend National Park from Marathon, while U.S. 385 northward goes to Ft. Stockton.

A turbid, fast-moving slurry of sediment and water, set loose by an earthquake, flows down continental slope and is deposited on deep sea floor. Much Haymond and Tesnus deposition was by turbidity currents. Each sand bed seen in roadcuts represents a separate turbidity current event.

Haymond formation.

Prominent above the alluvial fan and gravel surface north of Marathon is Iron Mountain, a distinctive, dark-colored, rough-looking peak of Tertiary intrusive rock; it stands out amongst the light gray, bedded sedimentary rocks of the surrounding ranges.

A veritable wall of limestone faces you in the first string of hills west of Marathon. North of the highway the westward-tilted ridges of Permian-aged rocks of the Glass Mountains attest to another great unconformity. These Permian rocks, 200 million years old, rest directly on the deformed older Paleozoic rocks, over 300 million years old, of the Marathon uplift. At the southern tip of the Glass Mountains is Cathedral Mountain, aptly named.

Caballos novaculite on ridge north of US 90.

Iron Mountain north of Marathon is a distinctive Tertiary intrusive body. Note dip to west of beds of sedimentary rock (Devonian limestones) in skyline ridge.

Cathedral Mountain northwest of Marathon is a ridge of tilted Permian limestones.

10 miles east of Alpine. Large cliff of Permian limestone (Capitan formation) and a small volcanic intrusion at the mouth of Ramsey Draw to the right of the photo.

Where U.S. 67 joins U.S. 90 about eight miles east of Alpine, Permian limestones are well exposed in an impressive cliff on the face of the mountain south of the highway. At the west end of the mountain face, see if you can spot the small volcanic intrusion at the mouth of Ramsey Draw. In the Draw and to the west of it are younger rocks of Cretaceous age, lying on the back side of the westward-tipped Permian section.

Two miles east of Alpine, igneous dikes are seen to cut young Tertiary volcanic rhyolite beds. By now, you have again crossed a major geologic boundary, having stepped from the territory of Paleozoic rocks in the Marathon uplift to much younger volcanic rocks of the Davis Mountains and Paisano Volcano.

From the town of Alpine you can look westward at the profile of the Paisano volcanic field outlined against the clear West Texas sky.

U.S. 285
Ft. Stockton—Pecos—New Mexico
106 miles

Between Ft. Stockton and Pecos, Highway 285 follows a northwest-southeast course across the gravel and sand surface of the Ft. Stockton Plateau. This sheet of sediment was spread southward by streams, mainly the Pecos River; the sediment came from erosion in nearby mountain ranges in New Mexico. Some of this sediment is probably also derived from the carving back of the caprock escarpment to the east, as the Pecos River chewed its way southward.

Dune sand is piled up in many places; the sand was blown out of the dry Pecos River bed and from the sandy gravel of the Ft. Stockton surface. Strong westerly winds frequently ply across the plateau, as seen in the stacks of tumbleweeds along fence rows. These winds blow the sand into dunes, commonly anchored by bushes and shrubs that act as baffles to trap the sand.

About the middle of this road segment the ground is quite flat; pebbles along the highway show the gravelly nature of the surrounding sediments.

Outcrops of older rocks stand above the general surface in two places along this road. About ten miles north of Ft. Stockton, low ridges and roadcuts are seen in a Cretaceous limestone ridge that

trends northeast–southwest across the highway. These middle Cretaceous limestones are similar in age to the thick Santa Elena limestone
seen in Big Bend National Park. Ten miles south of Pecos, another
ridge crosses the road, this one of Cretaceous limestones and colorful
Triassic shales and sandstones. Both of these outcrops are high-
standing ridges of rock that resisted erosion; the gravelly deposits
built around, not over them.

Between Pecos and the Triassic outcrops the road crosses Toyah
Lake, a windswept, dry, alkaline, white flat that hardly warrants the
name "lake." But after a rainstorm, the water collects here, and a
genuine lake exists until sun, wind, and evaporation return it to its
normal dry state.

If the day is clear, look to the west for a view of the skyline ridges
of the Delaware Mountains about 50 miles away.

Pecos—New Mexico Line

North of Pecos, U.S. 285 parallels the Pecos River, which is to the
east, but it is far enough away that you can't see it for most of the drive.
At the junction with Texas 302, you can look eastward in the Pecos
River valley, where the rolling topography indicates the river is not far
away.

The road continues to ride mostly on the flat, gravelly terrace
surface, but occasionally low skyline hills of sand dunes add variety
to the scene.

On a clear day, look for the profile of the Guadalupe Mountains and
Guadalupe Peak, the highest point in Texas (8,751 ft.) 70 miles to the
west-northwest. The Delaware Mountains appear on the skyline off to
the west. Note the termination of the range northward.

A picnic area is located 33 miles northwest of Pecos on the east side
of the road. Shortly thereafter is the town of Orla, and you can see the
town of Red Bluff in the Pecos valley to the east. Red Bluff Lake is a
dammed segment of the Pecos River that extends across the border
into New Mexico.

The town and lake are obviously named for the red outcrops of
upper Permian rocks, seen in the knobbly, erosional topography as
you approach the New Mexico border. The river has cut deep enough
to expose a northeast-southwest ridge of Permian rock, which forms
the eastward end of a large expanse of Permian outcrops. These
extend westward into the Delaware Mountains.

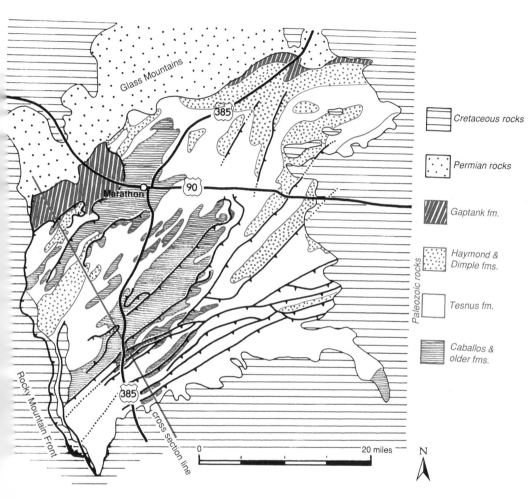

Geologic map of Marathon uplift.

The Marathon Uplift

In this remote, dry corner of Texas an almost magical aura surrounds a swirled-up tract of geology known as the Marathon uplift. Here, rocks stand on end, tipped vertically to point to the sky. Or, they gyrate around hillsides imitating pulled taffy; others glisten like white battlements in the rising heat waves above the stony desert. Nowhere else in the entire state of Texas are rocks seen so deformed, so bent, so crushed and folded and faulted as in the Marathon region. Travelers have the unique opportunity to peer into the deformed flanks of an old mountain range, which has been brought to the surface by intense crustal squeezing, then exposed by millenia of equally intense erosion.

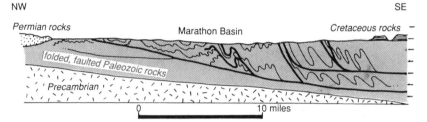

Cross section of the Marathon basin and uplift, showing the intensely folded and faulted Paleozoic rocks.

The Marathon uplift is one of only two exposed pieces of the buried Ouachita Mountain chain that arcs across Texas. The other is the nearby Solitario Dome located northwest of Big Bend National Park. As part of the Ouachita system, the Marathon region has a long and colorful geologic history. Throughout most of Paleozoic time, from Cambrian to Permian, or nearly 300 million years, a tremendous pile of sedimentary rocks, over 14,000 feet thick, was laid down, mostly in deep marine water.

These deep water sediments are intriguing because of the way they were deposited. Many of the sandstone and boulder beds in the Tesnus and Haymond formations, for example, were deposited by turbidity currents. These instantaneous bursts of water and sediment, triggered perhaps by an earthquake, flow at breakneck speeds down the steep edges of continents, only to finally deposit their sediment load in the deep sea at the base of the slope. The speed of some turbidity currents has been clocked at over 50 miles per hour!

Equally intriguing are the white, hard, pure silica rocks of the Caballos novaculite of Silurian-Devonian age, about 400 million years old. Novaculite, a term borrowed from the Ouachita Mountains of

Arkansas is used there to describe similar hard, dense, light-colored, sedimentary silica rock. Since the Marathon region is part of the Ouachita Mountain belt, it is appropriate to use the term here, and then to realize how widespread this type of rock is, as it extends from West Texas all the way to Arkansas. Because the novaculite is so hard, it resists erosion and is easily recognizable as white ridges over much of the Marathon uplift. Continents came crashing together about 300 million years ago in the Pennsylvanian to Permian period. The trough that had been receiving sediment for millions of years was arched upward, and the pile of sediment was pushed laterally for at least 125 miles to the northwest. In the process, the sedimentary rocks were pummeled into long northeast-southwest folds, while fault after fault cracked through the rocks.

Where a trough had been, a mountain range now stood at the end of Paleozoic time. But, what goes up must come down in the earth, and the forces of erosion began their inexorable chewing at the new range. In the short time of only a few million years, the range was worn low enough that seas again lapped over the surface; stacks of Permian and

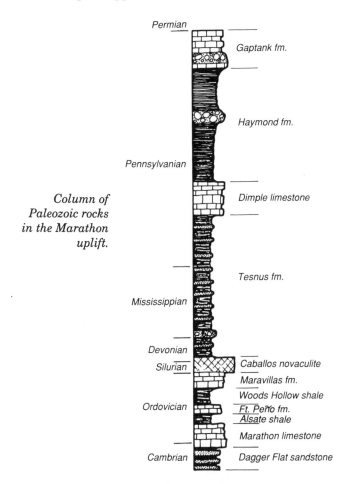

Permian

Gaptank fm.

Haymond fm.

Pennsylvanian

*Column of
Paleozoic rocks
in the Marathon
uplift.*

Dimple limestone

Tesnus fm.

Mississippian

Devonian

Silurian — Caballos novaculite

Maravillas fm.

Woods Hollow shale

Ordovician — Ft. Peño fm.

Alsate shale

Marathon limestone

Cambrian — Dagger Flat sandstone

287

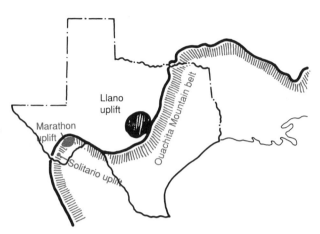

Marathon and Solitario uplifts are the only surface expressions in Texas of the Ouachita Mountain belt.

Cretaceous-aged limestones were laid over the area, burying the deformed Paleozoic rocks.

In its turn, this later realm of silent sedimentation came to an end as North America was set adrift to the west by a new pulse of seafloor spreading and continental migration. The Laramide mountain-building epoch (80 to 55 million years ago) had begun; a new northwesterly range of faults and uplifts pushed on the old Marathon structures from the west. The Rocky Mountains had arrived on the scene. And, once again, erosion moved mountains to the sea, grain by grain. Erosion removed the blanket of Permian and Cretaceous rocks to expose the old, underlying Paleozoic rocks. Further doming in late Tertiary time, 38 to 15 million years ago, was accompanied by the intrusion and extrusion of immense volumes of molten lava. The adjacent Davis Mountains came into being, along with smaller satellite plugs and dikes. Erosion, erosion, and more erosion from 15 million years ago to today, gave the landscape its characteristic mesas, hills, valleys, and exposed intrusions.

Roads crossing the Marathon uplift are U.S. 90 between Alpine and Marathon, U.S. 90 for twenty miles east of Marathon, and U.S. 385 between Marathon and Persimmon Gap at the northern portal of Big Bend National Park.

How can you be driving down to get to an uplift!?

The Marathon area is a structural uplift, because old rocks are elevated to the surface.

But topographically, the Marathon area is low, surrounded by higher, younger rocks. Erosion has removed the younger rocks to expose the older section.

U.S. 385
Marathon—Big Bend National Park
40 miles

This road segment illustrates the folded and faulted character of the Paleozoic rocks of the Marathon uplift and gives you a chance to view close-up the strange Caballos novaculite in roadcuts along the way.

Just after turning south onto U.S. 385, small roadcuts of 500 million year old Ordovician - Cambrian Marathon limestone are seen. South of town, on hills and ridge tops, the white gleam of Caballos novaculite is striking, particularly to the east on the Woods Hollow Mountains and on the long ridges to the west of the highway. Three miles south of Marathon the road cuts through a ridge of Maravillas limestone/Caballos novaculite. Note how the beds are tipped steeply to the south. Notice also how hard, dense, fractured, and sharp (be careful!) the siliceous novaculite is in the roadcuts. It is likely that the bodies of marine sponges were significant contributors of the silica in the novaculite, because tiny, distinctive sponge spicules can be seen in novaculite under a microscope. Silica needles are the only skeletal strengtheners in the body of sponges.

Caballos novaculite three miles south of Marathon on US 385.

Close-up of Caballos novaculite three miles south of Marathon on US 385. Penny (upper left) for scale.

Watch for a picnic area about nine miles south of Marathon. From here to the west you can see "flatirons" on East Bourland Mountain. Intricate folds and turns in the white Caballos/dark Maravillas outcrops on Horse Mountain (5,012 feet) to the east graphically illustrate the twisted-up, intensely deformed character of the Marathon uplift.

The road again slices the hard novaculite at 12 miles south of Marathon, where nearly vertical beds of rock are displayed in a nice roadcut.

Flatirons of Caballos novaculite on the mountains west of the highway, nine miles south of Marathon.

Tertiary volcanic intrusion and Cretaceous Glen Rose limestone mesa, 24 miles south of Marathon on US 385.

About 20 miles south of Marathon, the surrounding landscape undergoes a fundamental geologic change. To the east is the jagged profile of the Tinaja Mountains, where the hard Paleozoic novaculite holds up the ridge. But, looking south, the mid-Tertiary intrusion forming Santiago Peak looms on the skyline, and flat-lying Cretaceous limestones surround the road at 24 miles. We have just crossed from the erosional window that looks into the Marathon uplift and entered the non-eroded realm of Cretaceous limestone overburden, which is punctuated by later Tertiary intrusions. This latter package of rocks dominates the surrounding landscape on into Big Bend National Park.

Profile of Santiago Peak 20 miles south of Marathon on US 385 on the way to Big Bend Park.

BIG BEND NATIONAL PARK

Introduction

According to Indian legend, when the Great Creator had finished making the Earth, a large pile of rejected stony material was left over. Already finished with the main job, the Creator threw this material into one heap and made the Big Bend country.

Talk of a national park in the Big Bend country began when American troops were stationed in the area while Pancho Villa held sway in northern Mexico. Talk continued for 30 years until Big Bend

National Park was officially established June 12, 1944, and Dr. Ross Maxwell, a national park service geologist, was named first superintendent.

Big Bend park preserves unique and nationally significant natural phenomena: it contains the outstanding section of Chihuahuan Desert wilderness in the United States, with plants and animals occurring nowhere else. It is a mixing zone where Rocky Mountain species from the north meet Mexican highland species from the south. And, biologic zones climb from wet, moist floodplains of the Rio Grande, through vast tracts of dry Chihuahuan Desert, upward to the cool, moist elevations of the Chisos Mountains where pine forests predominate.

Historically the park is also rich and fascinating, emanating from colorful border towns, isolated ranches, mercury mining, and Indian lore. But it is the geology of Big Bend National Park that strikes the visitor in an overwhelming display of topography, odd erosional forms, volcanic remnants, fossil beds, and sheer cliffs of clearly exposed stratigraphy. This great geologic diversity, all painted in

Roads in Big Bend National Park.

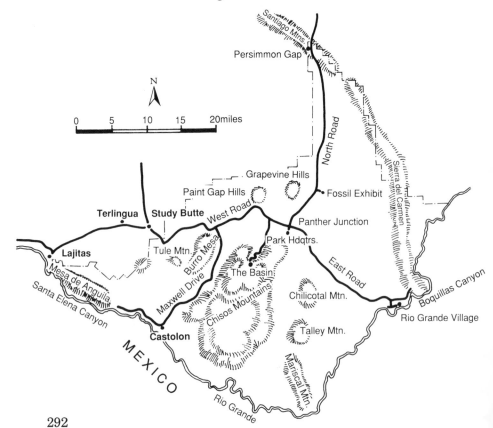

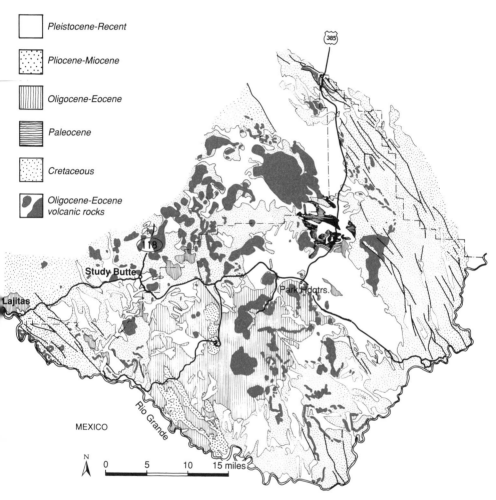

Pleistocene-Recent

Pliocene-Miocene

Oligocene-Eocene

Paleocene

Cretaceous

Oligocene-Eocene
volcanic rocks

385

118

Study Butte

Lajitas

Park Hdqtrs.

MEXICO

Rio Grande

N

0   5   10   15 miles

Geologic map of Big Bend National Park.

ever-changing colors, becomes even more captivating to one who wants to read the rocks for their meaning. Indeed, the rocks of Big Bend have amazing stories to tell!

Geologic Overview

Like the giant Colossus, Big Bend stands astride the junction of two major mountain ranges. Rocks of the Ouachita Range come to the surface north of the Park in the vicinity of Marathon. At Persimmon Gap, Paleozoic-aged rocks of Ouachita origin are thrust over Cretaceous limestones. The inexorable push of the Rocky Mountains, responding to plate-crunching motion of North America sliding over the Pacific, is dramatically preserved in the northwest–southeast-oriented faults, folds, and block mountains forming the Sierra del Carmen on the east and Mesa del Anguila on the west of the Park. Indeed, the Ouachita Mountains meet the Rocky Mountains at Big Bend!

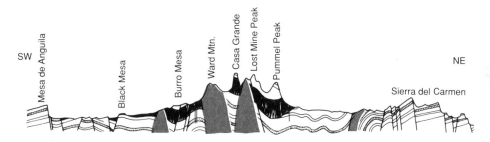

Cross section of Big Bend National Park.

But, after the elevation of the old Ouachita Range in Paleozoic time, about 300 million years ago, erosion of the range occurred and thick piles of Mesozoic sandstones, shales, and especially limestones were laid down over the old range. The elongate Cretaceous seaway, trending from the Arctic to the Gulf, right through Big Bend, was the site of tremendous piles of carbonate deposits made up mostly of the shells of sea-dwelling organisms. And then thick sections of stratigraphy were dramatically elevated thousands of feet above sea level when the new mountain-building episode at the end of Cretaceous time created the Rocky Mountains. The cliffs at Santa Elena Canyon and Boquillas Canyon tell this story of limestone deposition and subsequent uplift.

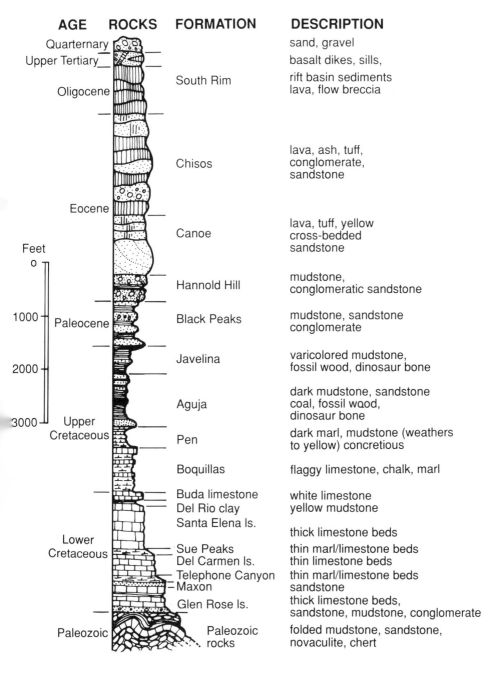

AGE	ROCKS	FORMATION	DESCRIPTION
Quarternary			sand, gravel
Upper Tertiary			basalt dikes, sills,
Oligocene		South Rim	rift basin sediments lava, flow breccia
Eocene		Chisos	lava, ash, tuff, conglomerate, sandstone
		Canoe	lava, tuff, yellow cross-bedded sandstone
Feet 0		Hannold Hill	mudstone, conglomeratic sandstone
1000 Paleocene		Black Peaks	mudstone, sandstone conglomerate
2000		Javelina	varicolored mudstone, fossil wood, dinosaur bone
3000 Upper Cretaceous		Aguja	dark mudstone, sandstone coal, fossil wood, dinosaur bone
		Pen	dark marl, mudstone (weathers to yellow) concretious
		Boquillas	flaggy limestone, chalk, marl
		Buda limestone	white limestone
		Del Rio clay	yellow mudstone
		Santa Elena ls.	thick limestone beds
Lower Cretaceous		Sue Peaks	thin marl/limestone beds
		Del Carmen ls.	thin limestone beds
		Telephone Canyon	thin marl/limestone beds
		Maxon	sandstone
		Glen Rose ls.	thick limestone beds, sandstone, mudstone, conglomerate
Paleozoic		Paleozoic rocks	folded mudstone, sandstone, novaculite, chert

Column of rocks in Big Bend National Park.

On the heels of Rocky Mountain compression came a period of apparent continental relaxation when western North America began to pull apart. Trenches opened up, normal faults groaned open, and following these new weaknesses in the crust, huge volumes of hot, molten, mantle material shot upward. The lava filled cracks, and it pushed upward in dome-shapes, finger-shapes, and mushroom-shapes, shoving at the overlying sedimentary rocks, sometimes reaching the surface to spit out froth, rocks, liquid, and gas in violent volcanic episodes that must have made Mt. Saint Helen's look like a birthday candle! Other vast amounts of hot rock never breached the thick limestone overburden, but caused subterranean blisters of various sizes and shapes, some even squeezing laterally between layers of the sedimentary column. This volcanic burst occurred quite simultaneously from Big Bend to New Mexico, to Colorado and even Montana in a wide belt, leaving behind vast volumes of lava, ash and debris. The Davis Mountains north of Big Bend Park are entirely built of volcanic material from this great episode.

Much of the topography in Big Bend Park is due to faulting in late Tertiary time that lifted and dropped great blocks of rock to form mountains and intervening basins or rifts. This activity in Big Bend Park is part of the broader pattern of Basin and Range extension that extended across the west. The central part of Big Bend is actually a graben or down-dropped basin called the Sunken Block.

The Sierra del Carmen on the east and the Sierra de Santa Elena on the west form the great blocks on either side of Big Bend's Sunken block. The main period of this extension and rifting was from about 26 to 10 million years ago. Grabens from this period are filled with sediment shed from adjacent highlands. One such late rift basin is seen in the southeast part of the park, where basin-fill sediments are exposed in eroded banks of Tornillo Creek.

But it is the story of erosion that finally molds Big Bend into its unique and peculiar morphology seen today. After all the tectonic pushing and shoving and breakage, after all the pyrotechnics of volcanic eruptions and deeper intrusions, the Big Bend relaxed and was quiet for millions of years. But, during this quiescent period the unending forces of water and gravity worked their incessant magic. The Big Bend has not always been so dry. When it was wetter, during the Ice Age (Pleistocene), trickles became rivulets, and muddy-wash streams turned into boulder-clunking torrents. Literally thousands of cubic miles of stones, rocks, and sand grains have been carried out of this country in the last few million years—a short period, really, in the cosmos of geologic time.

Think about erosion, then, as you stand in the Chisos basin and look 2,500 feet up to the top of Ward, Vernon Bailey, and Pulliam peaks. These are mountains built of intrusive igneous rocks. The body of this rock was intruded into the upper sedimentary layers of the earth's crust. These rocks cooled hundreds to thousands of feet beneath the surface, or, put the other way around, there used to be hundreds to thousands of feet of sedimentary rock on top of Ward, Vernon Bailey, and Pulliam peaks! Erosion has removed it all.

And, look around at all the other Big Bend intrusive bodies shown on the maps and discussed in this guide. Erosion has exposed them all. An enormous amount of rock has been weathered and moved and removed from this country!

Erosion has not worked with an even hand over the landscape, however. If it did we would see a flat surface. Instead, mountains, canyons, valleys, and mesas stand in unequal stature all across the face of Big Bend. Erosion works, but the rocks themselves have something to say about it too. Soft sediments erode quickly as their loose particles move even under the pattering of raindrops. Hard basalt layers, in contrast, mightily resist even the abrasion of landslide scraping or boulder-laden river pounding. But, not forever, and even hard basalt layers and tough limestone ledges eventually come tumbling down.

Keep this concept of "differential erosion" in mind when you look at mesas with hard tops and soft lower slopes, or see a high mound of hard intrusives above the surrounding landscape. The main tale of Big Bend topography can almost be written in one word: erosion. It is more fully explained in two words: differential erosion.

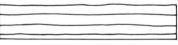

1 sedimentary rocks

Sequence of sedimentation, intrusion, and erosion in the Big Bend area.

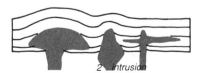

2 intrusion

3 erosion

If you would like more information about the geology of Big Bend, the book by Ross A. Maxwell, 1968, *The Big Bend of the Rio Grande: A Guide to the Rocks, Landscape, Geologic History, and Settlers of the Area of the Big Bend National Park*, published in Austin, Texas, by the University of Texas Bureau of Economic Geology, is recommended. The book is sold at Park Headquarters at Panther Junction and at the concession building in the basin of the Chisos Mountains.

The North Road
Persimmon Gap—Panther Junction
29 miles

The north Park entryway is in a low-elevation pass, called Persimmon Gap, through the northwest-southeast trending Santiago Mountains.

This pass has been the location of a road or trail for hundreds of years. It was first the Comanche Trail, then it was followed by early travelers, in turn by army camel trains, then by cowboys, freighters, Texas Rangers, stagecoaches, and ore wagons. Finally, travelers today quickly drive through the pass in air-conditioned cars on smooth pavement.

Stop at the Persimmon Gap Ranger Station to see geologic exhibits and maps of the Persimmon Gap area and Big Bend National Park.

The Santiago Mountains at Persimmon Gap have a somewhat complex, but fascinating structure. Paleozoic rocks, similar to those seen near Marathon, are thrust-faulted, older rocks over younger, in Persimmon Gap. This thrusting occurred during Ouachita Mountain building, about 300 million years ago. Then, during Rocky Mountain building time, about 70 million years ago, these rocks were again thrust-faulted, but this time over younger Cretaceous rocks: the Cretaceous rocks are internally thrust-faulted as well.

About one quarter mile north of the Ranger Station, Paleozoic rocks of the Tesnus and Maravillas formations are seen at the top of the hill. They were thrust over Cretaceous limestones of the Glen Rose formation, seen in the middle of the hill. These in turn are thrust over younger Cretaceous rocks of the Boquillas, Pen, and Aguja formations seen at the base of the hill.

All this pushing and shoving, at two periods of time, illustrates the complexity of the geology in this area. But, it also tells, in a rather

A thrust fault (fault 1) in Paleozoic time placed older Ordovician (Maravillas formation) rocks over younger Mississippian (Tesnus formation) rocks. The motion along this fault was toward the viewer. This faulted section was then folded along a later steep reverse fault (fault 2), which placed older lower Creteceous rocks over younger upper Cretaceous rocks during the post-Cretaceous, Laramide (Rocky Mountain building) orogeny.

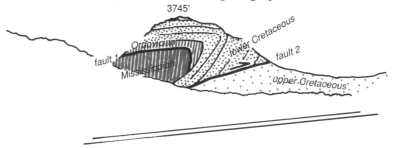

simple way, how the crust of the earth deforms to make mountains in response to the crushing forces of colliding plates. The horizontal motion of the plates squeezes rocks in the crust until they can't bear the strain anymore; finally they break. Low-angle thrust faults result and the final effect is the shortening of the crust. More rock is simply stacked up and squeezed into a smaller lateral space through the mechanism of thrust faulting.

South of Persimmon Gap, the ridges and walls of rock to the east are Cretaceous limestones, elevated in blocks along northwest–southeast-oriented normal faults. These ranges make up the Sierra del Carmen, which extends from Persimmon Gap southward to the Rio Grande.

To the west, Cretaceous rocks are punctuated by dark-brown, weathered, intrusive rocks of Tertiary age. Weathering has removed

Anticline in Cretacous rocks on Dagger Mountain, North Road in Big Bend National Park.

much rock from atop these hard intrusive bodies, and because of their hardness, they now stand higher than the surrounding countryside.

As you drive along, watch the Cretaceous rock walls to the east—spot the white-colored rock fall, where a big section of limestone gave away and tumbled to the talus slope below. Magnify this process thousands of times over thousands of years to understand how mighty mountains eventually come tumbling down.

The Rosillos Mountains predominate on the skyline to the southwest. Note the brown color and massive, rounded, but internally structureless appearance of these mountains, indicating the rocks are part of a large, intrusive igneous mass, one of but many such bodies in the Big Bend area.

Dagger Mountain is a large dome seen to the east. It is an anticline, where Santa Elena limestone beds have been arched up—you can see the limestone beds inclined all around the dome. Dagger Mountain was probably uplifted by an intrusive body, which is as yet unexposed by erosion.

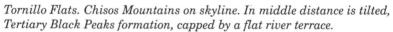

Tornillo Flats. Chisos Mountains on skyline. In middle distance is tilted, Tertiary Black Peaks formation, capped by a flat river terrace.

About 15 miles south of Persimmon Gap, the road enters colorful flatland and badlands terrain known as Tornillo Flats. Here are stream and flood plain sediments of brightly colored, purple, white, gray, and tan shales and sandstones that were laid down in early Tertiary time (Paleocene and Eocene). Particularly noticeable are flat terrace surfaces where erosion has not yet created "badlands". The colors and almost eery landforms in Tornillo Flats provide some of the most unusual and different scenery in the Park.

Colorful as they may be, these rocks are also famous for their treasure troves of fossil bones of primitive mammals that roamed these parts in early Tertiary time — 50 million years ago. You can see their actual remains in an exhibit constructed where the fossils were found. Watch for the sign to the fossil bone exhibit about 19 miles from Persimmon Gap, eight miles from Panther Junction. The side road heads eastward off the highway.

Fossil Bone Exhibit

The short drive from the highway to the exhibit parking lot brings you out into the Tornillo Flats scenery. Near the parking lot note the cross-bedded sandstones and pebble conglomerates that were deposited in stream channels. The varicolored shales were deposited in swampy, muddy areas between the channels. In this ancient environment lived the ancestors of the mammals—dawn horses, gazelle-like camels,

Early Tertiary mammals and environment in Big Bend about 50 million years ago.

mouse-sized animals and other creatures that look mammalian. But you can't quite put your finger on whose relatives they belong to in today's pantheon of familiar mammals. Their names are equally unfamiliar—*Ptilodus, Titanoides, Coryphodon,* etc. - but the illustrations of these animals in the exhibit are excellent and give you at least a "feel" for the fauna that lived in Big Bend 50 million years ago.

Back to the highway, and again heading toward Panther Junction, the Grapevine Hills lie off to the west. These Hills are the eroded remnant of a mushroom-shaped intrusion.

Nearing Panther Junction, the skyline of the Chisos Mountains is seen dead ahead and the little hill, standing alone to the west, next to the highway is appropriately named, "Lone Mountain." It is capped by a horizontal layer of igneous rock, a sill that injected laterally between the layers of Cretaceous Aguja formation. This hard rock is what "holds up" Lone Mountain by protecting the soft, surrounding Aguja shale from erosion.

Enter Panther Junction and Park Headquarters. In the building you will find books, maps, pamphlets, and displays on the geology, as well as the flora, fauna, and history of Big Bend National Park.

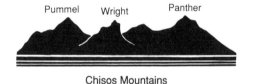

Chisos Mountains

Cross-bedded, Eocene sandstone at fossil bone exhibit.

Nugent Mountain.

The East Road:
Panther Junction—Boquillas Canyon
25 miles

Heading southeast, the road traverses alluvial fan deposits of sand and gravel that slope away from the Chisos Mountains. The gravelly fans were eroded from the Chisos Mountains and carried outward in bursts of watery flow in streams swollen by downpours following desert thunderstorms. Views to the south are particularly good of the upper volcanic crags—Wright, Panther and Pummel peaks—that make up the Chisos Mountains' distinctive skyline.

About four miles east of Panther Junction, and south of the road, Nugent Mountain stands out as an outlier of the Chisos Mountain complex. It is a hard mass of intrusive rock exposed to view by differential erosion: the softer rock surrounding the hard central mass has simply been more quickly eroded.

On the other side of the road, gentle, rather shapeless hills, and tan-brown sand and conglomerate deposits are seen—the product of stream deposits and later stream dissection, leaving high-standing terraces. Look for flat stream terrace surfaces at about mile 13.

Ahead on the skyline is the massive limestone wall of the Sierra del Carmen, so called in Spanish because of its red glow at sunset.

Prominent south of the road is Chilicotal Mountain. Chilicote is the Spanish name for the coral bean bushes that grow on the mountain.

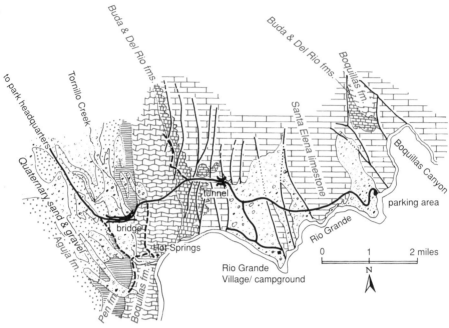

Geologic map of East Road; near Boquillas Canyon, the Tunnel, and Tornillo Creek Bridge.

The mountain is a body of intrusive rock, surrounded by shales and sandstones of Cretaceous age. Look also for the characteristic shape of Elephant's Tusk peak.

Where the river road turns off to the south, at about mile 16, roadside exposures of yellowish to brown Cretaceous sandstone of the Aguja formation display cross-bedding. Silicified wood and dinosaur bones have been found in this formation.

A mile farther on, the road crosses lower Tornillo Creek on the River Bridge. You might wonder why such a substantial bridge is built

Sandstone of the Aguja formation (Cretaceous) at junction of East Road and River Road.

Bridge over lower Tornillo Creek. Pen and Boquillas formations are backdrop to sand bars in dry river channel.

across this dry stream—flash floods are common in the desert following thunderstorm downpours—and water has actually flowed over the bridge here!

Just on the east side of the bridge is a spectacular change in geology as one formation contacts another. The yellow-weathered clays of the Cretaceous Pen formation are seen in sharp contact with the white limestone of the Boquillas formation, where abundant marine fossils of large clams are found. Look too, for a small vertical dike that cuts across the clay deposits a few hundred yards from the bridge.

Dike cuts Cretaceous Pen formation east of Tornillo Creek Bridge.

Tunnel through Cretaceous Santa Elena limestone.

The road traverses a mile or so of white limestone of the Boquillas. formation to the Old Ore Road, where white limestone gives way to brown, thicker bedded limestone of the Buda formation.

Hard, brown-weathered limestone beds of the Cretaceous Santa Elena formation form the hills around the highway, and are the rocks through which the tunnel up ahead is bored.

At the Boquillas Canyon junction, look north to see a very nice anticline in the limestone beds of the Santa Elena formation. All the limestone beds from the tunnel to Boquillas Canyon and Rio Grande Village are Cretaceous Santa Elena formation. You are also cutting

Patches of dune sand and wind ripples at mouth of
Boquillas Canyon.

across the Sierra del Carmen, a range composed of northwest–southeast-trending blocks of rocks impressively uplifted along linear faults that also trend northwest–southeast. Note that the road to Rio Grande Village is on terrace gravels deposited from an older stage of the Rio Grande, and that the village is located on the low-lying modern day flood plain of the river.

The road to the Boquillas Canyon parking lot crosses Santa Elena limestone for awhile, then passes some well-exposed roadcuts of stream gravels on the way down to the parking lot.

The trail hike to view the canyon mouth is well worth it. Look for round pot holes on the limestone bank of the river where Indians ground their corn. Sand banks farther on attest to higher levels of the Rio Grande stream flow at times. The awe-inspiring view of the Boquillas Canyon mouth from the cobble-sized gravel bar shows how powerful streams and their cobble-sized abrasives can be in cutting such a slash in solid rock given enough time. Look for the wind-piled stack of sand to the left of the canyon. Wind ripples are seen cutting across the face of the sand pile.

And, if you look closely you will see a nearly vertical fault offsetting beds of limestone inside the canyon on the right hand wall.

The Santa Elena limestones in Boquillas Canyon are the same rocks that form much of the cliff walls at Santa Elena Canyon on the other side of the park.

Entrance to Boquillas Canyon, where thick section of Cretaceous Santa Elena limestone is faulted (arrow).

The Basin Road:
Main Highway—The Basin
on Park Route 4
7 miles

The Basin Road turns south from the main highway three miles west of Panther Junction (Park Headquarters). Immediately the road begins to climb, traveling on the gravelled surface slope of the alluvial fan that splays outward from the Chisos Mountains.

On Pulliam Mountain, to the right of the road as you head south, look for the profile of Alsate's face—Indian legend says that when Alsate, an Apache leader, was killed, the earth shook and his face appeared on the side of the mountain.

Alsate's profile on Pulliam Peak in the Chisos Mountains.

Facing the Chisos Mountains, the rocks you see from bottom to top are mostly the product of igneous activity. The upper parts of the mountains to the left, east of the road are mainly extrusive igneous rocks - lava flows, ash deposits, and broken rock (breccia) that spewed out of a volcano about 40 million years ago in Eocene time. The lower slopes are composed of intrusive igneous rock that was pushed upward into overlying older volcanic and sedimentary rocks, but never got to the surface. This happened about 30 million years ago in Oligocene time. So, the intrusive rocks are actually younger than the overlying sedimentary and volcanic rocks that they penetrate. The mountains to the right, west of the road are massive bodies of intrusive igneous rocks. Pulliam Peak, Vernon Bailey, and Ward mountains in the Chisos Range are part of the same large, intruded igneous body, a "batholith."

As the road enters Green Gulch and you begin to see solid rock on either side of the highway, look for dikes that cut the rocks on the lower mountain slopes. You will also get a view of superb talus slopes

Green Gulch. Pulliam Peak to the far right, Casa Grande at center right, and Lost Mine Peak to the left.

forming skirts of eroded rock and sand and gravel on the lower slopes beneath steep cliff walls. Weathered cracks or "joints" give the brownish cliffs of intrusive igneous rocks the appearance of castles.

At about four miles down the road watch for the sign that tells you have just reached 5,280'—one mile above sea level. Casa Grande Peak—the biggest castle-like battlement, looms up ahead.

The road twists and winds upward through green shrubs and conifer trees, far different from the cactus, yucca, and ocotillo in the desert below. The apex of the road is at 5,800' at Panther Pass, which is the drainage divide between Green Gulch and The Basin ahead. The trail leading to Lost Mine Peak begins from here.

Vertical joints in intrusive rocks on Pulliam Peak in Green Gulch. Note steep talus slopes emanating from weathered joints.

WEST   NORTH   EAST   Casa Grande   SOUTH   WEST

The Window   Vernon Bailey Peak   Pulliam Peak   Toll Mt.   Emory Peak   Ward Mt.

campground   concession area & parking

Panoramic 360° view in the basin, Chisos Mountains.

After negotiating a couple of genuine hairpin turns, the downward trek of the road takes you into the heart of the Chisos Mountains, into The Basin. Gnarled and jagged peaks 2,500' above the basin floor surround you now at every point of the compass. With all the volcanic rocks around, you might think The Basin is a crater or caldera, but it is not. The Basin is the pure product of differential erosion, and the harder intrusive rocks of Pulliam Park, Ward, and Vernon Bailey mountains stand high. When the protective lava cover, seen as remnant on Casa Grande Peak, was penetrated by erosion, softer rocks below were swiftly eroded away and carried out of The Basin through the gap in the mountain facade known as "The Window." All the water and sediment carried off the slopes and peaks into The Basin go out through the window. Panther Pass, the low spot you drove over to enter The Basin is where Basin drainage and Green Gulch drainage are eroding towards each other and one day will create another outlet from The Basin.

The rocks on the west, The Window side of the Basin, are mostly intrusive igneous in origin; Pulliam, Ward, and Vernon Bailey peaks are all part of the same intrusive body. Extrusive lava flows, breccias and rhyolites form the peaks and slopes of Casa Grande, Emory, and

Casa Grande.

View in Chisos Basin. Ward Mountain to left, the Window to right, peak of Carter Mountain to immediate left of The Window.

Toll peaks. Talus on the lower slopes of Ward Mountain mostly conceals a band of Cretaceous limestone—Pen, Boquillas, and Aguja formations.

The volcanic stratigraphy of Casa Grande Peak is fairly well exposed and near enough to be seen. It is explained in the adjacent diagram.

The meaning of the word "Chisos" is an interesting one. It is generally accepted locally that the word means ghost, spirit, or enchanted, referring to the mystical aspect of the Chisos Mountains. An Apache word "chish-ee" means "people of the forest", and Indians living in the Chisos Mountains at the time of Spanish contact were mountain people. Maybe they were thus called "chivos," goat or mountain goat. Anyway, "Chisos" is probably the composite result of all these near-match words, reflecting the compelling history of these fascinating mountains.

Casa Grande Peak, Big Bend National Peak.

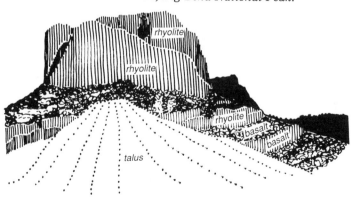

The West Road:
Panther Junction—West Park
Boundary and Study Butte
24 miles

At Park Headquarters, Lone Mountain to the north sits lonely above the alluvial fan plains. It is capped by a resistant igneous sill surrounded by Cretaceous rocks. Magnificent views of the Chisos Mountains are seen to the south, with Pummel, Wright, and Panther peaks forming the three highest crags—all part of the central pile of volcanic rocks that makes up the Chisos Mountains.

The first half of the road is on the sloping alluvial outwash emanating from the Chisos Mountains. At Government Spring, just beyond the turn-off to The Basin, the road crosses a mound of volcanic intrusive rock known as the Government Spring laccolith. (A laccolith is a mushroom-shaped body of lava intruded between the layers of overlying sedimentary rock. It cooled in place before being exposed by erosion.) The road then turns north to reveal a good view of the near-skyline ridge of the brown volcanic Paint Gap Hills. Croton Peak is at the west end of this range. The Rosillo Mountains, Corazon Mountains, and Christmas Mountains, east to west, form the distant skyline to the north. Past Croton Springs Road is a lonely little sentinel of yellow ·

Chisos Mountains, from the West Road, a few miles west of Panther Junction. Ocotillo in flower in foreground.

Tule Mountain looking south from the West Road.

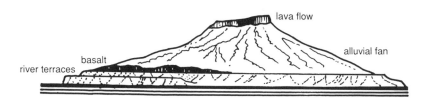

sandstone (Aguja formation) over gray shale, standing to the right of the road as a reminder of erosion. At the Maxwell Drive Junction, Burro Mesa to the south shows how lava flows "hold up the mesa," by forming a resistant layer to erosion.

The Burro Mesa fault lies at the edge of the mesa, separating volcanic rocks from Cretaceous rocks to the east.

Dissected terrace and fan deposits, cut by stream erosion, produce badlands seen to the north of the road, just west of Maxwell Drive junction. Nearby Dogie Mountain is a volcanic intrusion.

Look carefully to the southwest for fine views of Tule Mountain and Santa Elena Canyon as you drive along the flats west of Burro Mesa. Tule Mountain looks like a ship's prow, and is an obvious local landmark. The rhyolite layer at the top of Tule Mountain is the same age and composition as the rhyolite on Burro Mesa and Casa Grande Peak. Maverick Mountain, another volcanic intrusion, is the large mountain to the north of the highway.

Dead ahead is the formidable wall of Sierra Ponce/Mesa de Anguila. Santa Elena Canyon is the gap in the wall through which the Rio Grande flows.

Eroded valley-fill sediments. Note flat terrace surface on dissected alluvial outwash, and knobby volcanic rock in background.

To the north and west of the highway, knobby, chaotic, disjointed-looking volcanic peaks stand above dissected fans and badlands, colorfully exposed in hues of yellow, white, and tan in contrast to the dark brown volcanic rocks above. The lower colorful rocks are part of the Cretaceous-aged Javelina and Aguja formations, in which dinosaur bones and petrified wood are preserved.

After passing the old Santa Elena Canyon Road (Maverick Road) junction, the highway drops down into the badlands where Dawson Creek has cut into colorful Cretaceous-aged clay beds. Notice the terrace levels—flat tops—and erosional forms in the badland terrain.

The road climbs out of the badlands near Study Butte (pronounced "Stoody," named after a local doctor), an old mining town. Note the red and black mine tailings piled north of the highway at the outskirts of town. Unusual yellow-gray-white "hoodoos" of erosional origin are seen east of the central town junction. These contrast with the dark brown, upheaval appearance of the volcanic rocks north of town.

Maxwell Drive:
West Road—Castolon and
Santa Elena Canyon
30 miles

The turnoff to Maxwell Scenic Drive from the main highway (West Road) is 13 miles west of Panther Junction (Park Headquarters). This scenic route is named to honor Dr. Ross Maxwell, a geologist and first superintendent of Big Bend National Park. The Park Service has done a great job of explaining the geology along this road with numerous turnouts and illustrated signs.

Heading south, the flat-topped mountain to the right, west, is Burro Mesa. Along the sharp, top edge is a resistant bed of rhyolite, a lava flow that also appears on the top of Tule Mountain to the west, and caps Casa Grande Peak in The Basin of the Chisos Mountains. Vertical cracks in the face of this ledge are cooling cracks—as lava cools, it shrinks, and the cracks appear in the hardened flow. At the base of the steep slope-face is the Burro Mountain fault along which the mesa was uplifted.

East of the highway the Chisos Mountains dominate the skyline, and Ward Mountain is unmistakable out in front. Ward Mountain is part of a large intrusive rhyolite body that forms the western half of the Chisos Mountains.

Beneath Ward Mountain are a series of dikes that stand vertically and cut across the horizontal bedding of the Chisos formation, produc-

Syncline in Chisos formation below Ward Mountain.

315

Dike in roadcut extends into the distance forming a 'Chinese wall.'

ing a "Chinese wall" effect as they snake across the topography. Look for them crossing the road at about four miles from the main highway; a turnout and signs explain their geology.

A spectacular view of the very colorful stratigraphy on Burro Mesa is seen to the north of the road after you make the hairpin turns into the valley. The hard, brown-weathered rhyolite lava flow still caps the mesa. Below that, orange, red-brown, yellow, tan, and white beds of ash, tuff, lava, conglomerate, and breccia tell the tale of a series of volcanic eruptions, explosions and lava flows that must have over-whelmed the landscape about 35 million years ago in Eocene time.

Turning southward the mountain ahead to your left is Goat Mountain, and a display and turnout explain the geology. One of the most interesting aspects of the rocks exposed on Goat Mountain is that you can see an irregular surface of actual topography where an ancient stream flowed southwestward toward Castolon cutting a 900-foot-deep canyon! And later, into this canyon flowed volcanic tuff, lava and breccia, filling the canyon to the rim and spilling over the top.

Volcanic rocks weather to a brilliant array of colors on south end of Burro Mesa, just north of hairpin turns in the road.

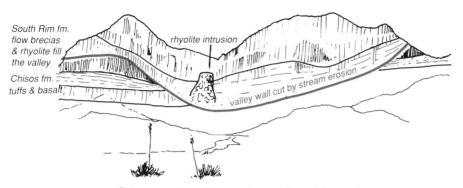

South Rim fm.
flow brecias
& rhyolite fill
the valley

Chisos fm.
tuffs & basalt

rhyolite intrusion

valley wall cut by stream erosion

Interpretation of the geology of Goat Mountain.

As you drive along, watch to the southeast for the distinctive twin-pronged Mule Ears Peaks. Early travelers looked for these landmarks to guide their travel. The peaks are a composite of volcanic tuff and intruded-lava dikes that erosion has molded into the shape of mule's ears.

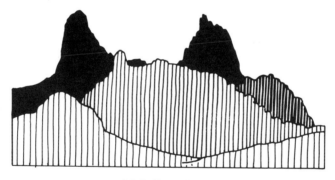

Mule Ears Peaks.

Nineteen miles down the road from the Goat Mountain turnout is another geologic turnout and display, at Tuff Canyon. A short walk takes you to an overview, where you will see bright white tuff, ash, and fragments blown explosively from a volcano. These rocks lie on dark lava, material that flowed out of a volcano. The contrast is distinctive and striking. And, of course, the power of erosion has exposed these rocks in the canyon walls. Note how the stream has cut easily through the soft tuff, but then met the hard lava, which resists erosion and forms ledges and flat, less erodable surfaces.

Tuff Canyon. Brilliant white volcanic ash over dark lava.

A few more miles along, the ash and lava deposits seem to close in on the road, and you drive through tight places where the rocks can almost be touched out the car window. White ash and black lava are interwoven in a stark landscape of artfully eroded shapes. Watch carefully along the right for what looks like a tree trunk partially exposed on a white slope. It has even been interpreted as a petrified tree trunk, but really is a small volcanic spine, a little volcanic neck. The hole on the side is not a branch, but a cavity where a pumice ball has weathered out. Things in this country are sometimes not what they first appear!

White tuff and dark lava. Small dark volcanic spine to the right of center looks tree-like up close.

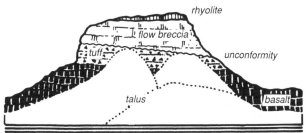

*Cerro Castelan
photograph and
interpretation.*

Nearing Castolon, Cerro Castelan, another distinctive landmark for early travelers and settlers, comes into view. The bright orange, red, and white breccia, basalt, and ash deposits on the lower slope contrast with the brown-weathered rhyolite — the same unit as we've seen on Burro Mesa — that caps the hill. Cerro Castelan is held up by a volcanic spine.

The road drops down onto gravelly river terrace deposits as it enters Castolon Village. The village was named after a settler who lived nearby. Lt. Echols of the U.S. Army visited Castolon in 1860, and the army had a garrison here during the border troubles of 1914 - 1916. The trading post is an old cavalry barracks, the two houses were army officers quarters, and the Park Ranger's office was once a Texas Ranger Station.

From Castolon, the road parallels the Rio Grande on the way to the Santa Elena Canyon overlook. You pass the abandoned houses of an early settlement called Coyote. Sandstone ridges of Cretaceous Aguja formation are seen on the right side of the road. The rock wall of Sierra Ponce looms to the left on the Mexican side of the River.

About eight miles from Castolon is the Santa Elena overlook parking area. A display in a small shelter at the edge of the parking area has geologic exhibits that explain the geology of Sierra Ponce, Mesa de Anguila, and Santa Elena Canyon.

319

The rock wall through which the Rio Grande has cut its gorge is an uplifted block of Cretaceous limestones. The Terlingua fault is at the base of the massive cliff. You are standing on the downthrown side of the fault, and the uplifted block facing you represents 3000 feet of movement along this fault!

The canyon separates Sierra Ponce in Mexico from Mesa de Anguila in the U.S. The long, thin Mesa de Anguila was likely named for its shape (anguila = eel), or for the freshwater eels found in the river here, though there is also the suggestion anguila is an English misspelling or corruption of the words aguila (eagle) or angulo (angle or corner).

Terlingua Creek, coming from the right (north) flows along the face of Mesa de Anguila and joins the Rio Grande at the mouth of Santa Elena Canyon.

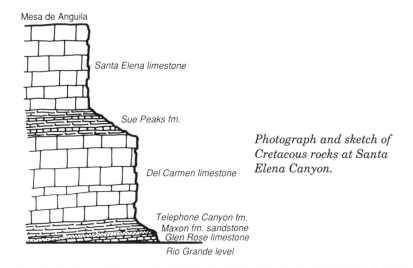

Photograph and sketch of Cretacous rocks at Santa Elena Canyon.

An obvious question is why the Rio Grande cuts across this massive uplifted block of limestone—why doesn't the Rio Grande just go around it on the way to the Gulf of Mexico? The answer is that the Rio Grande (or its ancestor) was here before the block was uplifted! And, as the block rose along the fault in earthquake increments of a few inches to a few feet at a time, the river simply downcut a little faster to accommodate the small change in slope caused by the last earthquake and uplift. Now, repeat this process many, many times over several million years, and you can see how the river cut Santa Elena Canyon little by little, inch by inch—nothing dramatic, just small effects multiplied over millions of years.

Texas 17
Ft. Davis—Marfa
21 miles

The northern half of the drive is across desert flats amongst volcanic mountains that are continuations of the Davis Mountains complex. Blue Mountain is the first large range seen to the west, after leaving Ft. Davis driving south. It is mainly composed of 35 million year old Barrel Springs volcanic rocks. The Puertacitas Mountains are visible to the southeast on the skyline. Knobby hills on either side of the road about ten miles from Ft. Davis are basalt lava flows. The large dome to the east at mile ten forms the west end of the Puertacitas Mountains. The road goes over a low pass here as it crosses the end of this range. Looking back, or if you are driving north look ahead, to the skyline ridges of the Davis Mountains.

In Alamito Creek, two miles north of Marfa, are beautiful, small-scale terraces. Take a good look at their shape and form to better understand how streams meander back and forth as they cut downward, leaving behind these flat ledge-like terraces.

The south half of the road is in Marfa Basin country. Stream and gravel fill from surrounding ranges makes up the floor of the basin.

The town of Marfa was named by the wife of the chief engineer of the Southern Pacific Railway in 1882. She was reading Dostoevski's 1880 book "The Brothers Karamozov" and gave the Karamozov household servant's name, Marfa, to this 1880's railroad watering stop and freight headquarters.

Texas 54
Van Horn—U.S. 62-180—
Guadalupe Mountains National Park
65 miles

Traveling north on Texas 54 out of Van Horn, the distinctive skyline profile of Three Mile Mountain is seen to the west. The mountain has a protective cap of hard Permian Hueco limestone,

Three Mile Mountain looking west from Van Horn.

whereas the lower slopes are outcrops of extremely ancient sandstones and conglomerates deposited before most life forms were developed on Earth. The Van Horn sandstone here is Precambrian, more than 600 million years old! These are some of the oldest rocks in the state. To the east is a broad flat valley, called the Salt Basin, which is a down-dropped segment of crust between two uplifted blocks. In this low area windblown sands and white alkali salt flats are prevalent. The hills seen in the distance to the east are the Apache Mountains, part of the east-side uplifted block.

About five miles north of Van Horn, Texas 54 closely skirts Beach Mountain, seen close-by to the west. This mountain is a domed and faulted block of Ordovician and Cambrian rocks, though plenty of old Precambrian rocks surround the mountain on its west flank. The road passes Beach Mountain and then turns abruptly west to get between Beach Mountain and the Baylor Mountains to the north.

The Baylor Mountains are also a faulted block, capped by Permian Hueco limestone. Ordovician and rare Silurian rocks are exposed on the eastern flank. Silurian rocks are not preserved nor seen in outcrop

Beach Mountain. Upper ridge is (Ordovician) Montoya dolomite, lower slopes are thin-bedded sandy dolomites of the (Ordovician) El Paso formation.

Baylor Mountains, looking north. Note inclined Ordovician rocks at left.

Sierra Diablo, looking west. Magnificent talus slopes skirt the lower slopes.

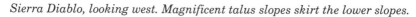

very commonly throughout the western United States, so these are special outcrops to many geologists interested in the Silurian period.

After passing through the narrow gap between Beach Mountain and the Baylor Mountains, the road turns north and flanks the high, sheer, impenetrable front wall of the Sierra Diablo Mountains for the next 25 miles. What a range to cross! No wonder it was named Diablo, the devil!

The Permian Hueco limestone is the high cliff face at the near-top of the range; you can trace it northward for many miles. Beneath it are beautiful talus slopes that form "skirts" for the mountain. The talus nearly covers the thousands of feet of older, Precambrian Van Horn and Hazel formation sandstones and conglomerates. These reddish sandy beds are nicely exposed, however, at the very south end of the Sierra Diablo.

Follow the limestone ledge northward, and notice that it gets lower and lower; as you reach the mouth of Victoria Canyon, the Hueco limestone is at road level and younger Permian rocks are at skyline. The Sierra Diablo is an uplifted block, but it is also tipped toward the north!

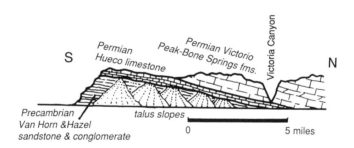

Southern end of Sierra Diablo. Permian limestone (Hueco formation) forms a high wall. Red Precambrian sandstone and conglomerate beds are nicely exposed below, but are covered northward by talus.

About 20 miles south of U.S. 62-180, 35 miles north of Van Horn, the face of the Sierra Diablo takes an abrupt turn to the west away from the road; on the corner of the range, low down on the slopes, another set of rare Silurian-aged rocks is exposed. Apache Canyon gouges through the range just beyond the corner.

The road from here to the junction with U.S. 62-180 traverses the middle of the Salt basin. To the east, the Permian strata in the Delaware Mountains are magnificent. These long, horizontal layers of sandstone, along with limestone and shale can be traced for miles. They were deposited into the sea at the edge of the Delaware basin. Some of the shale beds were laid down on the seaward slopes below the great Permian reef that is now exposed on the cliffs in Guadalupe Mountains National Park.

About five to seven miles south of the junction with U.S. 62-180, the road crosses low hills where sandstones of the Permian Brushy Canyon formation rise above the salt flats.

As the road nears U.S. 62-180, views to the north get better of the imposing cliff walls of El Capitan and Guadalupe Peak in the Guadalupe Mountains. Guadalupe Peak is the highest point in Texas, rising to an elevation of 8,751 feet above sea level.

Guadalupe Mountains National Park

Inaugurated as a National Park in 1972, largely through the conservation efforts and land donations of a renowned petroleum geologist, Wallace Pratt, Guadalupe Mountains National Park preserves the largest fossil reef in the world.

The horseshoe-shaped reef, nearly 400 miles long, formed during Permian time, about 250 million years ago, around the edge of the Delaware basin, which was an arm of the Permian ocean that covered west Texas and southeast New Mexico. Portions of this reef are also exposed in the Apache Mountains near Van Horn, and in the Glass Mountains near Marathon, Texas.

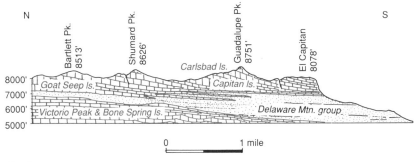

Geologic cross section of Guadalupe Mountains. Note how the inclined limestone beds, which formed the reef front, tongue into the deeper marine fore-reef deposits.

El Capitan. Reef rocks of the Capitan limestone form the peak. Below are fore-reef sandstone, claystone, and limestone beds of the Delaware Mountain group; Cherry Canyon, Bone Spring, and Brushy Canyon formations.

Modern reefs are found in shallow, warm, clear ocean water, built by the growth of lime-secreting, marine organisms, mainly corals. Similar conditions in the sea around the Permian basin probably existed but the principal reef-builders were lime-secreting algae, not corals. Sponges, bryozoans ("moss-animals") and clam-like brachiopods were the other contributors to the Capitan Reef. The limy skeletons of these organisms piled up through millions of years, creating a formidable carbonate wall in the Permian sea. The reef built upward to a height of 1,300 feet, but also migrated seaward for miles. The reef's seaward face was pounded by waves through the eons, loosening pieces of reef rock that fell as debris in front of the reef, creating a wide talus debris apron. The organisms grew out over the talus, extending the realm of the reef seaward.

But this was not a continuous process, because at times sea level fell, leaving the reef high and dry, killing the organisms, only to rise again. The animals then repopulated the reef to start the cycle over. These cycles of the reef's life can be seen in the different colored portions of the high walls on the Guadalupe escarpment.

On the backside of the reef, a shallow lagoon trapped sediment from streams, and limestone, mudstone, and sandstone built up the floor of the lagoon. The lagoon waters commonly evaporated, leaving a soupy brine too salty for most marine organism. But it was ideal for the formation of vast expanses of evaporative minerals such as anhydrite and gypsum, along with algae-laden limestone.

Finally, the connection between the Permian basin and the open sea was permanently cut-off and the basin water evaporated, leaving behind the thick deposits of salt and potash. Over more millions of years the low areas filled up with sand and silt brought in by streams, and the old reef, lagoon, and basin itself were buried under thousands of feet of sediment.

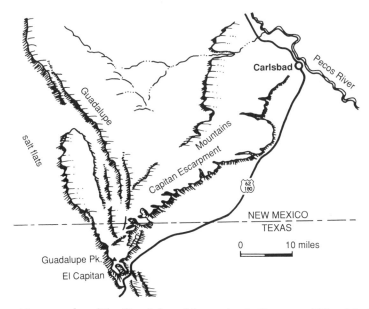

Present topography of the Guadalupe Mountains in Texas and New Mexico.

Uplift that began with Rocky Mountain building about 70 million years ago was enhanced during the late Tertiary when regional extension, about 10 million years ago, further lifted the deeply buried Permian reef. As uplift occurred, erosion relentlessly removed the sediment that once was piled over the reef. Moreover, as the reef rocks were uplifted they fractured, and as erosion cut downward toward the reef, ground water from the surface found its way into the fractures within the reef limestones. The slightly acidic ground water dissolved the limestone over time and created numerous caverns, including the colossal Carlsbad Caverns to the northeast.

The attack by erosion continued until the sedimentary cover was entirely removed from the reef. The hard limestone resisted further

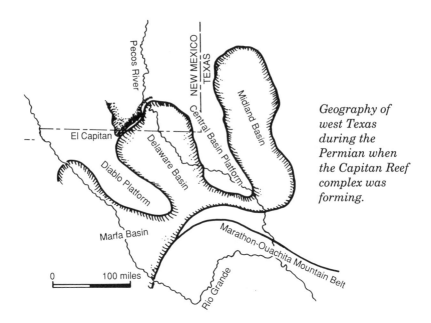

Geography of west Texas during the Permian when the Capitan Reef complex was forming.

erosion, while the softer rocks surrounding it were removed, until today the reef stands tall above the deep-water rocks of the Delaware basin in a near mirror-image of the Permian reef-talus-basin scene of 250 million years ago. It looks almost as if someone had simply pulled the plug on the sea!

The upper cliff faces of El Capitan are the limestone reef rock, while the lower shaly and thin-bedded limestone slopes are the deep water sediments deposited in the basin in front of the reef. Large blocks of limestone debris, representing talus deposits, can be seen in lower slopes if you look very carefully. The best views of the reef structure in the Guadalupe Mountains are seen from the west looking eastward at the escarpment. Look for the steeply inclined beds in the upper limestone cliffs that represent various stages of the steep reef front.

A fine view of El Capitan can be had from the scenic turnout along US 62-180 on the Guadalupe Pass entry road. Outcrops along the road cut through flat-bedded limestone, sandstone, and shale beds of the Delaware sequence that were deposited in deep water in front of the reef.

The Frijole Visitor Center and McKittrick Canyon Visitor Center have geological displays. The Permian reef geology trail leaves from the McKittrick Canyon Center.

For a nice review of the history, geology, and ecology of the Park, you may want to buy the booklet, *Trails of the Guadalupes, a Hiker's Guide to the Trails of Guadalupe Mountains National Park*, 1986, by Dan Kurtz and William Goran; published by Environmental Associates, Champaign, Illinois. It is more than just a trail guide.

Texas 118
Study Butte—Alpine
79 miles

The southern half of the road is on Cretaceous, 80 million year old limestone terrain, punctuated by a large number of Tertiary intrusive volcanic rock masses.

The northern half traverses Eocene, 30 to 40 million year old volcanic rocks that are part of the gigantic Davis Mountain volcanic field which extends from the Rio Grande 130 miles northward nearly to Interstate-10. Watch for blue roadside signs that give names and elevations of the significant peaks along the way.

Coming north out of Study Butte, the road for about ten miles is literally studded on either side by intrusive masses that have punched upward through the surrounding Cretaceous rocks. Erosion has left them exposed as hills and mountains.

East of Study Butte is Maverick Mountain, two miles north of town is Bee Mountain on the west and Indian Head Mountain east of the highway, and four miles out of town is Willow Mountain standing east of the road. All three are typical examples of intrusive rock bodies in this area. Willow Mountain is particularly striking because of its vertical rock joints. The mountain mass is an intrusive plug, which cooled beneath the surface and in the process contracted, forming vertical, uniformly spaced cooling cracks or joints. The mass has been

Willow Mountain four miles north of Study Butte. This intrusive plug cooled to form spectacular vertical joints.

Packsaddle Mountain 14 miles north of Study Butte. Inclined, light-colored Cretaceous beds surround a dark-colored central intrusion.

Santiago Peak.

Crossen Mesa capped by basalt flow.

exposed by erosion, and weathering has removed soft rock, enhancing the joint pattern. Willow Mountain is probably the best display of rock joints in Big Bend country.

Continuing north, watch for low mesas of thin bedded, flat-lying Cretaceous rocks near the road. About 14 miles from Study Butte, look for the Packsaddle Mountain sign (elevation 4,661 ft.). The mountain is cored by a dark-colored volcanic intrusion, but you can readily see light colored, upturned beds of lower Cretaceous rocks on the flanks that were bent at a high angle as the mass pushed upward.

Camel's Hump, to the east, is aptly named. It is a two-humped camel! North of Camel's Hump is an obvious anticline in Cretaceous rocks.

At 20 miles, the large mesa to the east is Nine Point Mesa, capped by a Tertiary sill of lava about 1,000 feet thick.

The large dominant peak to the east, a landmark seen for many miles along Highway 118, is Santiago Peak, a large intrusive mass (elevation 6,521 ft.).

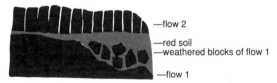

—flow 2
—red soil
—weathered blocks of flow 1
—flow 1

Sketch and photo of roadcut 23 miles south of Alpine, where two lava flows are separated by a red soil zone.

Profile of Cathedral Mountain (6,860 ft. elevation).

View to Alpine from high road on Big Hill south of town.

Phenocryts are common in brown weathered, dark gray-green basalt in roadcut on Big Hill south of Alpine.

At 50 miles north of Study Butte (30 miles south of Alpine), and after crossing rather flat terrain, with a few low roadcuts in tan, upper Cretaceous, flaggy limestone and marl beds, a high, dark, flat-topped mesa appears to the northwest. This is Crossen Mesa, capped by a 35-million-year-old, columnar-jointed lava flow.

The landscape from Crossen Mesa northward to Alpine is of low, rolling hills exclusively on volcanic terrain. Basalt flows, ash-deposits, and intrusions can be seen in mountains and roadcuts along this stretch. For example, at mile 56 from Study Butte (23 miles from Alpine) is a wonderful roadcut on the west side of the highway, where two dark lava flows are separated by a bright red soil. You can even see weathered blocks of the lower flow mixed into the red soil.

The distinctive profile of Cathedral Mountain to the west shows up for miles. It is another intrusive mass, exposed by erosion for all to see.

On a high hill, Big Hill, about nine miles south of Alpine you get a good view of the town to the north. Also, on Big Hill are excellent roadcuts of dark gray-green basalt that weathers brown. Note the large crystals scattered amongst the finer groundmass. From here, the road descends into Alpine, which is the home of Sul Ross State University, where a fine small museum has displays on the geology and history of the area. The town seems appropriately named, as it stands 4,481 feet above sea level at the entrance to the Davis Mountains, commonly called "The Texas Alps."

Mitre Peak.

Texas 118
Alpine—Ft. Davis
26 miles

The first few miles north of Alpine, the road is on a flat, alluvial plain. Off to the left (west) are mesas composed of Oligocene volcanic flows. Mitre Peak is an obvious landmark, shaped like a bishop's mitre hat. It is an intrusive mass exposed by erosion, and stands tall because it is harder than the eroded rock around it. A flat mesa, seen to the north where the road turns west, is capped by a dark brown ledge of Oligocene rhyolite and ash flow tuff.

As the road leaves the flatland to weave its way among eroded volcanic hills, look for the intrusive peak of Barillos Dome on the left (west). The road swings around the Dome to give you a good three-dimensional view of the central intrusive core surrounded by up-turned younger rocks on the flanks.

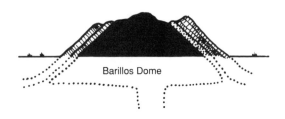

Barillos Dome

Lava rests on ash deposit. Note bubbles at base of lava, and darker, baked top of ash. Roadcut at south end of Musquiz Canyon.

Just as the road enters Musquiz Canyon, examine closely the large roadcuts, where you will see a dense, hard lava flow resting with a knife-edge contact on a gray, punky ash bed. Note that the top of the ash is *baked* from the heat of the overlying lava. Gas bubbles are seen at the bottom of the lava bed, where gas escaped toward the bottom edges of the flow as it began to cool. This is really a classic bit of geology!

Lava flows create characteristic flat mesas. Their associated talus slopes indicate much weathering is accomplished by gravity in this desert country, as rocks break off cliff faces and tumble down.

Columnar-jointed lava flow caps mesa. Talus slope below is composed of tumbled blocks of lava.

*Cliffs behind Old Fort Davis are 36 million year old Sleeping Lion forma-
tion, a rhyolite ash flow tuff with well-developed columnar jointing*

The landscape of intrusive bodies is different: note the typical rounded shapes of intrusions, with their chaotic joint and weather patterns.

At Ft. Davis, the cliffs behind old Fort Davis stand like a medieval castle as a backdrop to the fort. The cliff is a 36-million-year-old rhyolite ash-flow tuff with well-developed columnar joints. The tuff goes by the colorful name of "Sleeping Lion formation."

Texas 118
Ft. Davis—Kent (I-10)
50 miles

This road segment traverses the heart of the Davis Mountains, called by some "the Texas Alps," though you won't see any peaks that look anything like the Matterhorn. The Davis Mountains are, however, a gigantic volcanic area that had its origin 35 to 39 million years ago when western North America erupted in a violent spasm of volcanism. Volcanic rocks of Eocene to Oligocene age are found in a wide belt from Mexico to Montana, and the Davis Mountains are one of the largest volcanic centers in this chain. All kinds of volcanic strata make up the Davis Mountains—lava flows with their columnar jointed palisade appearance and widespread, thick ash-flow tuffs are most common. Dark, iron-rich lavas, basalts, are present, but are thin and not-so-common. The magma that formed these rocks either flowed out or was blasted out of two main volcanic centers—the Paisano Volcano west of Alpine, and the Buckhorn Caldera northwest of Ft. Davis.

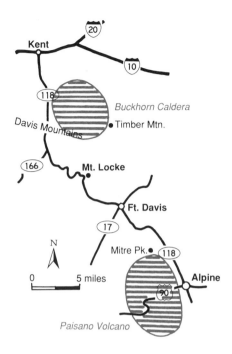

Map showing two large volcanic centers in Davis Mountains area.

As you drive across the Davis Mountains on Highway 118, note the many layers of horizontally-bedded volcanic rocks that comprise this vast eruptive system. The scenic hills surrounding the road are erosional imprints over the flat volcanics—in other words, you are *not* seeing individual volcanic cones, but rather a system of hills and valleys cut through the flat volcanic layers by stream erosion.

North of Ft. Davis, Highway 118 passes Ft. Davis National Historic Site before turning westward into Limpia Canyon a mile out of town. The brown cliffs of ash-flow tuff of the Sleeping Lion formation above the old fort continue into the canyon and can be seen after you make the turn westward.

At the entrance to Davis Mountains State Park are pink and red roadcut exposures of weathered gray rhyolite, also Sleeping Lion formation. Look for big feldspar crystals floating in a matrix of very fine-grained rhyolite. This texture reveals that the magma was already starting to crystallize at depth, but was then erupted to the surface, where the rest of the melt cooled rapidly.

Around the parking lot near the campground in the State Park are large blocks of local rocks where you can closely examine fresh rock faces and see the tremendous variety of colors, textures, and compositions of Davis Mountains volcanic rocks.

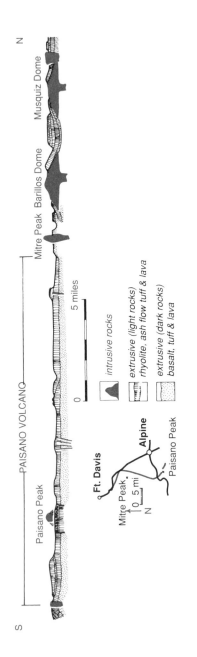

Cross section along Texas 118 between Alpine and Fort Davis.

Indian Lodge in Davis Mountains State Park. Grassy slopes and pinnacles of dark weathered volcanic rocks surround the lodge.

Rolling hills, valleys, tree-covered slopes and very pleasant vistas are seen for several miles. Occasional volcanic outcrops pop out on some hillsides and in a few roadcuts, but the vegetation on the weathered, rounded slopes covers much of the rock. Rainfall in the Davis Mountains is much higher than the surrounding desert country. The mountains cause westerly winds to rise over them; as the air climbs it cools and drops its moisture. The temperate vegetation reflects this increased moisture supply, and contrasts sharply with the desert plants in the lowlands surrounding the Davis Mountains.

The road climbs the flanks of Mt. Locke where the University of Texas McDonald Observatory is located. Several of the astronomical domes can be seen from the highway. The largest one belongs to NASA, the next largest one houses the 107 inch McDonald telescope. The visitor's center is open daily, but you need reservations for the public telescope viewing on the last Wednesday evening of each month. Call (915) 426-3423.

Below the observatory, where the road slices diagonally across the flanks of Mt. Locke, is a wonderful roadcut, complete with its own parking area and scenic overlook. A lava flow exposed in the cut is part of a bold natural outcrop that slashes across the mountain flank. The flow's sharp basal contact rests on ash-flow tuff, turned red in a soil zone, weathered at the surface between the time of the deposition of ash flow and the lava flow.

Texas 118 roadway cuts through lava flow on Mount Locke below McDonald Observatory.

Close view of white siliceous dike cutting lava flow. White quartz walls form the baked edge, with punky tuff center. McDonald Observatory roadcut.

Dark lava over red soil. McDonald Observatory roadcut.

Spheroidal weathering.

Exfoliation.

Zoned feldspar crystals. Penny for scale.

Near the overlook-end of the roadcut is a vertical, white silica dike that cuts the lava beds. Note how the walls of the dike are hard, very glossy silica, whereas the center is a bit coarser and more crystalline. The walls cooled fast, while the center cooled somewhat slower and had a little time to form larger crystals.

Watch for weathering effects in roadcuts west of Mt. Locke: you can see spheroidal weathering where jointed blocks of lava weather to form piles of rounded "boulders" that are not really boulders. You can also see exfoliation where layers of rock weather to form an "onion-peel" effect. Some of the outcrops also display large feldspar crystals, many of which are "zoned." The magma first crystallized feldspar in a magma chamber below ground, then erupted and chilled to form the fine-grained surrounding rock.

The road descends toward the west and the junction with Texas 166. From here northward to Kent, note the low topography and desert vegetation in contrast to the grass and trees higher on the Davis Mountains.

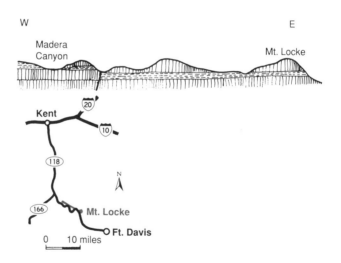

Cross section in Davis Mountains showing how widespread layers of volcanic tuff, rhyolite, and trachyte are eroded into hills and valleys.

Travelling north, the road passes through a canyon-like stretch between mesas of dark volcanic rocks, where a series of 37 million year old Oligocene lava flows stack up to nearly 1,000 feet thick.

As the road approaches Kent and Interstate 10, yellow-tan, thin-bedded, flat-lying beds of upper Cretaceous limestones, siltstones, and mudstones are seen on surrounding low mesas. These Cretaceous rocks were here before the Davis Mountain lavas pushed up through them and flowed out over them, forever changing the landscape of West Texas.

Texas 170
Study Butte—Terlingua—
Lajitas—Presidio
67 miles

This road is one of the most scenic, if not the most scenic, drives in the entire state of Texas. Dark volcanic mountains brood over lush-green copses of riverside trees, while blazing white, intricately carved ramparts stand guard over their Rio Grande moat. The road soars over high passes to offer grandiose views, then plunges to river depths where enveloping cliffs reveal their innermost geologic secrets. And, barely clinging to life like a tenacious desert plant, is the almost-ghost town, Terlingua, once the mighty producer of a fourth of the country's supply of mercury.

Between Study Butte and Terlingua, the road weaves through excellent roadcut exposures of yellow-tan, flaggy beds of mostly limestone, the Boquillas formation, laid down in shallow marine sea water in late Cretaceous time. The road crosses Terlingua Creek,

Upper Cretaceous Boquillas formation rocks along Texas 170 near Terlingua.

Terlingua. Perry Mansion, ruins, and General Store.

which flows south along the fault in front of Mesa de Anguila and empties into the Rio Grande at the mouth of Santa Elena Canyon. Both the mesa and canyon can be seen on the southern skyline, whereas to the north, peaked hills of dark volcanic intrusive bodies contrast sharply with the light-colored, layered limestones at road level.

It is hard to miss the mine tailings, abandoned stone houses, and old mine workings around the mercury-mining town of Terlingua. Turn in and visit the ruins and general store, still in operation. Besides soft drinks and souvenirs, displays of old mining equipment, mercury flasks, and a large painting of Terlingua in its mining days are on display.

Mining for mercury began at Terlingua in 1894 and continued under Chisos Mining Company auspices from 1902-1946. Howard E. Perry was the owner from 1898-1942 (his mansion is preserved in Terlingua), until Brown and Root Construction Company took over. During its heyday, 2,000 people, mostly Terlingua residents, worked in the mines and processing plant. The Terlingua mines were one of the most important sources of mercury in the United States, accounting at one time for a quarter of the nation's production. Mercury has many uses, but fulmonate of mercury in wartime blasting caps and bullet primers was the chief end- product of the liquid metal. Mercury is also a catalyst for making chlorine, and was used in thermometers, switches and gold extraction. However, as wartime applications have decreased, and as environmental knowledge has shown mercury to be a serious pollutant, being poisonous to humans, animals and plant life, the uses of mercury today are very limited. Consequently, as the ore above the water table played out and mercury use decreased, mining activity at Terlingua shut down in the early 1970s.

Oligocene-Eocene Chisos formation on Lajitas Mesa at town of Lajitas. The Chisos is a complex assemblage of sedimentary conglomerates and mudstones, and volcanic flows and tuffs.

The mercury mineralization at Terlingua occurs as red veins of cinnabar, the mineral name for mercury sulfide, which cut through the surrounding limestone beds and volcanic rocks. Hot ground water, carrying the mercury in solution, penetrated upward through cracks, faults, and fissures, and as it rose the solution cooled, leaving behind the cinnabar precipitate in fractures and small cavities in the limestone.

The crushed ore was heated in a furnace to release pure mercury vapor, which was then liquified in large condensers in a process similar to the way water vapor condenses on a cold windowpane. The mercury was collected and sold in flasks, each weighing 76 pounds.

The gray-weathered solid limestone cliffs of the Reed Plateau, south across the highway from Terlingua, are Cretaceous in age, just a bit older than the thin-bedded Boquillas limestone at road level.

Between Terlingua and Lajitas, the road continues to cross various limestone units of Cretaceous age. After passing white, thick beds of steeply-dipping Cretaceous limestone of the Pen formation in roadcuts about four miles west of Terlingua, you will see the roadside topography open out into a broad valley where huge alluvial fans slope toward the river. Intrusive volcanic rocks are seen to the north as odd-shaped hills and mesas, whereas Mesa de Anguila and its battlement walls of solid Cretaceous limestone fortify the sky to the south. Imposing flatirons of tilted Cretaceous limestone are also seen along this stretch, where they border the south flank of a large fold called the Terlingua monocline.

Tilted sections of flaggy-bedded, yellow-tan Boquillas limestone still greet you around the town of Lajitas, located at the base of Lajitas Mesa, where Comanche Creek joins the Rio Grande.

Lajitas is developing as a small tourist resort, and the new museum and desert garden east of town is a pleasant stop. Across the Rio Grande at Lajitas notice how the large mesa surface curves downward toward the river giving the distinct impression that rocks can bend, fracture, and break, if pressure is applied slowly, rather than quickly, in the earth's crust.

A fundamental geologic change takes place at Lajitas. Between Study Butte and Lajitas, the dark-brown volcanic piles and mounds are eroded *intrusive* bodies which drove up into, but not through, the surrounding sedimentary rock cover. West of Lajitas, however, the volcanic rocks seen from the roadside are mainly *extrusive*: they were either blasted from or flowed from a volcano. On the flanks of Lajitas Mesa, north of the town of Lajitas, are wonderful exposures of dark lava flows and white ash deposits. And, the road west of Lajitas traverses spectacular scenery carved from these volcanic emanations, now 15 to 30 million years old.

In roadcuts within a few miles west of Lajitas you can see white ash beds lying on dark lava flows. Small faults that offset individual layers of ash are quite clear in these exposures.

Small fault in ash beds in roadcut one-half mile west of Lajitas.

White ash over dark lava bed, one mile west of Lajitas. Air bubbles at top of lava indicate gas was being released to the surface as flow cooled.

Five to six miles west of Lajitas, the road swings northward in a loop away from the river. At the head of the loop, where the highway dips down to cross the bridge over Fresno Creek, flaggy, light-tan yellow limestone beds of the Boquillas formation stripe the creek's vertical walls.

Look up at the mountain to the northwest to see a spectacular face of alternating dark lava and white ash. Also note the offset of the black

Upper Cretaceous Boquillas formation in streambed by low bridge at the head of big loop on Texas 170.

Gravity slide which has displaced a section of dark lava flow.

lava layer, caused by gravity faulting and collapse of the front of the mountain. This is *real* geology in action!

Equally spectacular eroded ash deposits are to be seen along the road next to the river a few miles west of the Fresno Creek loop. You can make out all sorts of imagined faces, figures, and animals in these weathered forms, appropriately called "hoodoos" by geologists. An up-close look at this outcrop reveals holes and pits where crystals, softer than the surrounding ash, have been removed by the selective etching of sparse rain water.

On the cliff above the teepee picnic area, a lava flow displays very nicely-formed columnar joints: as the lava solidified, it shrank, and

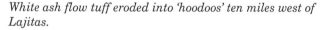

White ash flow tuff eroded into 'hoodoos' ten miles west of Lajitas.

Talus slope below vertical, columnar jointed lava flow at Teepee picnic area, 12 miles west of Lajitas.

these contraction cracks developed from the top to the bottom of the flow.

The road climbs over a high pass on the flanks of Santana Mesa to get around a steep, narrow canyon of the Rio Grande, creating a vantage point for marvelous views of the ragged volcanic scene to the west. An intrusive volcanic body flanks the east side of the big hill, but extrusive flows and ash make up the rest of Santana Mesa.

View west from high point on Texas 170 to Rio Grande and surrounding volcanic terrain.

An almost endless variety of shapes of eroded lava hills are seen west of Santana Mesa, as the valley becomes wider and wider. For the last third of the way to Presidio, the highway traverses immense alluvial fans which slope from the northern skyline mountains toward the Rio Grande. Cienega Mountain, a sizeable Tertiary intrusion, is seen to the north, and ahead in the distance beyond Presidio are the Chinati Mountains, formulated from several intrusions and piles of 30 million year old Oligocene extrusive rocks.

Old Fort Leaton at Presidio speaks of border battles in the tumultuous times of the early part of this century.

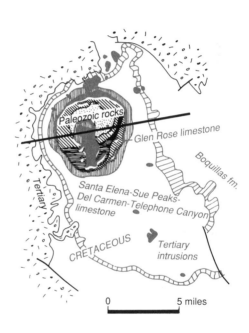

Map and cross section of Solitario Dome, showing the uptipped circular rim of Cretaceous rocks, the rim sill, the central peak of volcanic breccia and collapsed Cretaceous rock, and eroded center of dome where contorted Paleozoic rocks are exposed. The implied mechanism is upheaval by an igneous intrusion (laccolith).

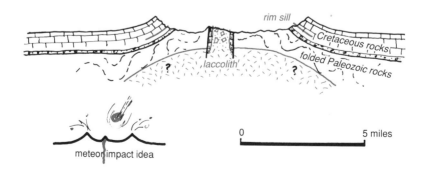

Solitario Dome

In the desert terrain northwest of Terlingua is a remarkably circular feature, eight miles in diameter, called the Solitario Dome. A 400-foot-high rim of upturned Cretaceous limestones isolate the Solitario from outside view, which, in addition to its remoteness, probably accounts for early Spanish settlers naming this fascinating geologic feature "El Solitario." The dome has rocks unlike any in the immediate vicinity, for the center is an eroded window into contorted, folded, and faulted Paleozoic rocks. These are identical to the Paleozoic rocks found in the Marathon uplift located north of Big Bend National Park.

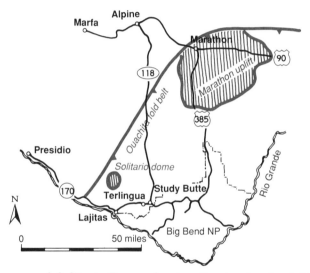

Location map of Solitario Dome, showing its structural relation to the Marathon uplift and Ouachita fold belt.

The origin of the Solitario Dome has been argued by some geologists to be a meteor crater, not an extraordinary interpretation considering that other meteor craters, such as Meteor Crater near Odessa and Sierra Madera north of Marathon, occur in West Texas. Other geologists, and probably the majority, believe the Solitario Dome was uplifted by a magmatic intrusion, a laccolith, in Eocene to Miocene time, about 20 to 45 million years ago. This uplift brought folded rocks of the old Ouachita mountain belt near to the surface, where later erosion exposed them in the core of the dome. The Solitario thus contains the most southwestward exposures of contorted fold-belt rocks of the Ouachita mountain system in the United States. The other localities are in the Marathon uplift and in the distant Ouachita Mountains in Oklahoma.

The Solitario Dome, now located in the new Big Bend State Natural Area, is an important piece of the geologic puzzle of West Texas, and distinctively marks the trend of the Ouachita fold belt.

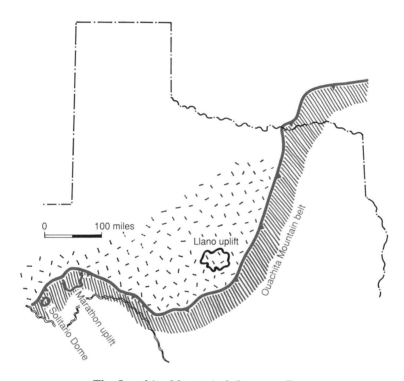

The Ouachita Mountain belt across Texas.

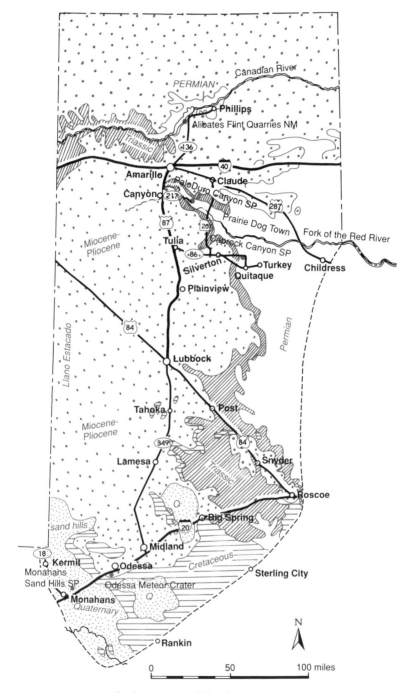

Geologic map of Northwest Texas.

VII
Northwest Texas
The Panhandle—High Plains

The high flat, windswept plains of the Texas Panhandle are seemingly not the "stuff" of which geology is made. But the flat surface and canyoned-borders of the high plains have a fascinating geologic history of their own, despite our prejudices that "real geology" is somehow concerned only with tall outcrops on precipitous mountain slopes!

The Texas High Plains, also known as the Llano Estacado, or "Staked Plains" in Spanish—so called because the early Spanish explorers laid out stakes across the plains to find their back trail—or so the story goes,* are high (average elevation over 3,000 feet) and flat (not more than a few tens of feet variation in topography). Midland and Odessa are at the south end of this plateau that nearly covers the Panhandle, and extends up to the northern Texas border with Oklahoma and over to the western border with New Mexico. Though the surface is flat and "featureless," it is not horizontal. In the northwest corner of the Panhandle the elevation is 4,000 feet near Dalhart, but slopes eastward and southward, and near Big Springs is 2,400 feet above sea level.

To understand the genesis of the Texas High Plains we have to look to geology beyond the state borders. In Miocene time, about ten million years ago, the Rocky Mountains to the north and west of Texas underwent uplift; erosion of these ranges increased dramatically, causing a vast apron of gravel, sand, and clay to be built eastward from

*The "Staked Plains" tale is deeply entrenched in Texas mythology, but the real interpretation of Llano Estacado is sensibly geologic: it means "stockaded" or "palisaded" plains—which is precisely how the edge of the plains appear when viewed from below the caprock!

the mountain front. This wedge of sediments once ran continuously from Wyoming, southward through eastern Colorado and into the Texas Panhandle Country. The so-called "Gangplank" sloped eastward from the mountains, as streams and rivulets spread their erosion load of sediment farther and farther eastward.

For many years geologists interpreted the main sediment body of the Gangplank as a series of coalescing alluvial fans and called it the Ogallala formation*. The formation name is a famous one on the High Plains because the Ogallala is the chief freshwater aquifer in the region. However, modern sedimentological investigations have led today's geologists to interpret the early Ogallala sediments as a series of braided stream deposits which filled pre-Ogallala valleys in humid climatic conditions. Then, as the climate progressively became sub-humid to arid, thick eolian (wind-blown) sand and silt covered the earlier fluvial deposits. Finally, caliche layers developed to cap the Ogallala, reflecting continued arid conditions with sparse rainfall, much like today's climate on the High Plains.

During the last two million years, the Pecos River eroded headward to capture the eastward flowing streams, which cut the drainage across the Panhandle, and left Llano Estacado, a high standing plateau, isolated from its former source, the Rockies.

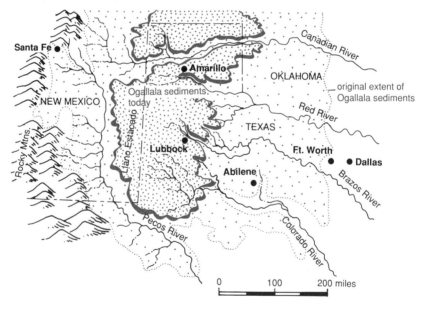

*Sediments of the Ogallala formation were deposited between ten and four million years ago, or during the last half of the Miocene epoch and first million years of the Pliocene epoch. Note on the geologic time scale that the Miocene ranges from about twenty-four to five million years and the Pliocene from five to one and a half million years ago.

This cycle of erosion and deposition and spread of sediment from the mountains continued until the Ice Age or Pleistocene epoch began. With vast amounts of water on the earth tied up in continental glaciers that covered much of Europe and North America during the Ice Age, sea level correspondingly fell about 400 feet. Rivers found their mouths not only lower but relocated seaward tens to hundreds of miles. At the same time the Pleistocene was a period of high rainfall, and formerly small streams became raging torrents. Since rivers adjust their gradients in response to the volume of sediment load and water supplied to them, in balance with the distance they have to travel, the Pleistocene rivers adapted and began to downcut across the Gangplank.

The Canadian River cut an impressive canyon across the northern panhandle to form the colorful "Breaks" country. Small streams eroded headward, and the edge of the High Plains was moved little by little westward. Many streams that today seem too puny to effectively carve the large canyons through which they flow carried more water in wet Pleistocene time. The Prairie Dog Fork of the Red River is an example: it hardly seems capable of carving the spectacular Palo Duro Canyon southeast of Amarillo. But it did when it carried more water! And, of course, this small stream still has tremendous erosive power when its water volume increases many fold after stormy downpours.

Given the flatness and high rainfall, much water was ponded at the surface, and a characteristic feature of the Texas High Plains is innumerable round ponds (commonly called 'playas' or 'playa lakes'). When the weather is dry, they are dusty, round, gray, usually unvegetated flats, as observed from the highway. But after a High Plains thunderstorm, water quickly fills the ponds, only later soaking into the underlying porous sandstones just below the surface to add to the groundwater in the Ogallala aquifer.

Early pioneers depended dearly on water from these surface ponds for themselves and their livestock, considering how few streams are on the High Plains. But rains didn't always come, and the ponds dried up frequently.

The 20th century has witnessed a concerted effort to tap the more reliable Ogallala water sands. Predictably, the consequent high dependency on groundwater has removed more water than is naturally replaced, raising concern for Panhandle citizens and planners as to future water supplies.

Southwest of the High Plains is a large area of active sand dunes that have been blown against the western escarpment by the strong westerly winds that frequently sweep across the southern Panhandle.

Along I-20 in the midst of the dune field, is Monahans Sandhills State Park, where sand dunes are available for study and play.

The eastern escarpment of the High Plains runs roughly 200 miles north-south from Amarillo to I-20. Road logs in the section cover several of the main roads that cross this topographically intriguing and colorful strip.

Far to the north, Alibates Flint Quarries National Monument and Meredith Lake Recreation Area are found in "The Breaks" country where the Canadian River has etched its way through the High Plains. This, too, is colorful country and worth a visit if you are in the Amarillo area.

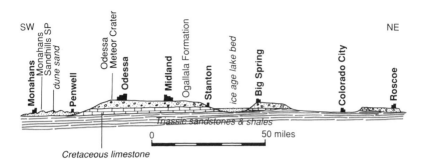

Cross section along Interstate 20 between Monahans and Roscoe.

Interstate 20
Monahans—Odessa—Midland—
Big Spring—Roscoe
154 miles

The area east of Monahans is a gigantic sand dune field. Tan sand is seen for miles on either side of the highway, piled in dunes, and trapped by vegetation, or on windy days, blows across your windshield in pelting streams of abrasive particles! Sand has been blowing and piling and abrading here since the Pleistocene (about 1 million years ago). The sand source is from the west, out of the sands brought down from the adjacent New Mexico mountains by the Pecos River.

Six miles east of Monahans is Monahans Sandhills State Park, where the State of Texas has preserved a beautiful segment of dunes for public use and appreciation. The entry and headquarters are right next to the Interstate, with easy highway exit and entry. You can get off the Interstate and spend a few minutes to a few hours enjoying the dunes. A geologic description and photos of Monahans State Park are found at the end of this road section.

If you are driving east on I-20, as you approach the town of Penwell look ahead to see the skyline profile of the High Plains. The edge of the Caprock escarpment is off to the south, which demarcates the southern end of the High Plains. Against this ridge face, held up by flat-lying Cretaceous limestone, the dune sand has been piled over time.

359

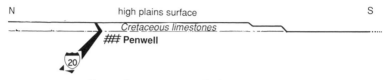

N   high plains surface   S

Cretaceous limestones

Penwell

(20)

Caprock escarpment skyline looking east.

The Interstate gently climbs up this face. Though no outcrops are seen near the highway, several quarry operations are seen where limestone is mined to use for construction gravel and road building.

On top of the High Plains, the flatness of the surface is apparent all the way to Odessa and Midland. It is dry here, vegetation is sparse, and the surrounding countryside is home to three kinds of critters: jackrabbits, pumpjacks, and tough cattle.

White rubble in the fields is about all the geology you'll see to tell you of the vast expanse of caliche deposits and gravel of the Miocene-Pliocene Ogallala formation that everywhere underlies the surface of the High Plains. And, a few tens of feet under the Ogallala is another vast expanse of flat-lying Cretaceous-aged limestone that you saw near Penwell on the edge of the High Plains escarpment. These same Cretaceous limestones peek from beneath the Ogallala surface in a band that crosses I-20 at the town of Stanton, 20 miles east of Midland.

In Midland, the Permian Basin Petroleum Museum has a large number of displays on the development of the petroleum industry in this oil-rich region of Texas. The history of early life on the southern High Plains is also presented in this outstanding museum.

Watch for signs to Odessa Meteor Crater located about five miles off I-20, southwest of Odessa. If you've got time, get off the interstate and eat lunch next to one of only two documented meteor craters in Texas. It is nothing like Arizona's spectacular meteor crater, but it's a real one, nevertheless. A description and photos of the crater are found at the end of this road section.

East of Stanton nearly to Big Spring, I-20 crosses a topographically low area that was once the site of an Ice Age lake bed; it formed when water and rainfall were much more abundant than they are today on the High Plains. East of Big Spring, I-20 is again on the High Plains surface of Ogallala sand and caliche. In 15 miles the road drops over the rather subdued eastern edge of the High Plains onto red and vari-colored beds and soils of the Triassic Dockum group, exposed by the headward erosion of upland tributaries of the Colorado River.

A narrow belt of Cretaceous limestone is encountered about ten miles west of Roscoe; these rocks underlie the edge of an isolated remnant of High Plains, separated by erosion from the main body of the High Plains. Flat scenery is seen again around Roscoe.

MONAHANS SANDHILLS STATE PARK

Six miles east of the town of Monahans, along Interstate I-20, Monahans State Park preserves a delightful tract of sand dunes for study and pleasure. All the features of sand dunes and their processes can be seen here; and it is just plain fun to climb and slide on the dunes!

The Monahans dunes are representative of a wide swatch of sand dunes that virtually cover the Pecos River area from Fort Stockton northward to the New Mexico border and eastward through Kermit and Monahans.

A ready supply of sand has been available since the Pleistocene Ice Age, 1 million years ago, in the eroded, dry flats of the Pecos River valley. Prevailing westerly winds blow the sand steadily eastward out of the Pecos valley, only to drop the sand grain by grain.

A large-scale factor in dune formation in this area is not easy to observe right in the park, but is important nevertheless. The Monahans group of dunes lie in a topographic depression, whose east, downwind, side is the escarpment of the High Plains, the Llano Estacado. As westerly ground winds rise to climb over the plateau, they lose velocity and drop sand.

Picnic area and dune field.

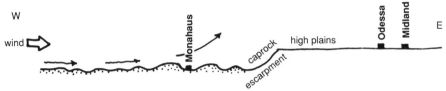

Wind drops sand as it rises over Caprock escarpment.

On a smaller scale, notice how the desert shrubs act as baffled barriers, trapping sand in piles behind them on the downwind side to form copse dunes. And once a sand pile forms, a continuous wind and sand supply will keep adding to the pile until it becomes a large dune.

Plants act as baffles to trap sand.

Large dunes grow as sand marches up the dune's windward side, only to fall as sandy rain on the steep side. Inclined wedges of sand are thus added to this downwind side resulting in large-scale "cross-bedding," a distinctive feature of dunes preserved in many ancient dune deposits.

On the ground, sand is moved by wind in two ways. The wind agitates the surface of dry sand grains, and they start to bounce around. A grain bounced up into the air flow will be pushed along in the air for a few inches until it settles back down, striking another grain, which in turn is bounced up. This process is called 'saltation,' and the length of the bounce depends on wind speed and grain size. The low, straight-crested ripples formed by this process can be seen

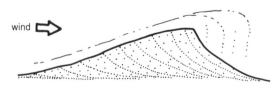

Dunes have large internal cross-beds.

Copse dunes—sand trapped by plants.

Wind ripples on surface of a transverse dune. Wind blows from the right.

everywhere in the park. The agitation of the surface also causes grains to continuously edge forward in a second motion aptly called creep. And, while your head is down looking for ripples, search for the tracks and trailways of insects and small animals amongst the rippled dune surfaces.

Movement of sand by saltation.

Dunes come in different shapes depending on wind conditions and sand supply. Two main types are seen at Monahans Park. Transverse dunes are long ridges oriented perpendicular to a single prevailing wind direction where sand supply is high. Barchan dunes are "C" shaped sand piles formed in a single prevailing wind where sand supply is low. Note that the 'horns' of Barchan dunes point downwind. Barchans appear commonly on the outer edges of larger dune fields.

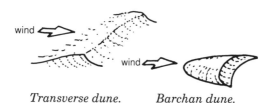

Transverse dune.   *Barchan dune.*

Other odd-shaped dunes, some in star shapes and snake-like patterns, occur in other dune areas of the world where wind patterns shift constantly, or blow from different directions at different times of the year.

The pleasant headquarters building, adjacent to I-20, houses a museum with exhibits on animals, plants, geology, and human history of the dune area. The Park is open 8:00 a.m. to 10:00 p.m. A small fee is charged. Lots of picnic areas and campsites are located amongst the dunes.

An excellent pamphlet on the detailed geology and history of Monahans Sandhills is by Marcie D. Machenberg, 1984, entitled *Geology of Monahans Sandhills State Park*, Texas, Guidebook 21," from the Bureau of Economic Geology, University of Texas, Austin, 78713.

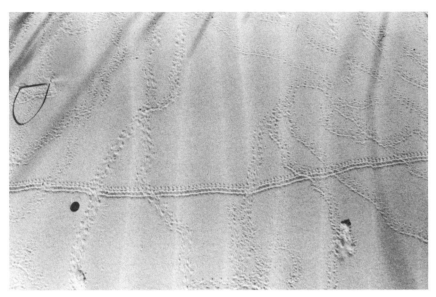

Insect tracks across wind ripples. Dark spot in the lower left is a penny for scale.

Lag surface of small pebbles in area between dunes. Saltating grains shift sand until pebbles are evenly distributed on surface.

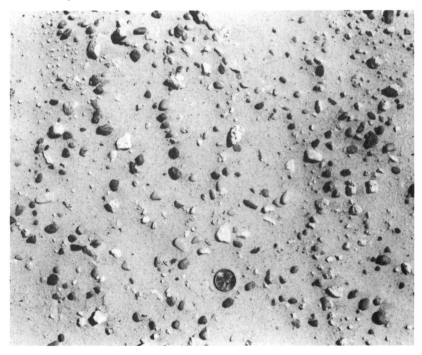

ODESSA METEOR CRATER

About five miles southwest of Odessa is the 550-foot-diameter Odessa Meteor Crater. Geologists believe a nickel-iron meteorite in a meteor shower emanating from the asteroid belt between Mars and Jupiter formed this crater about 20,000 years ago. The energy generated by the impact blasted a hole into the Cretaceous limestone bedrock of the High Plains surface and explosively shattered the surrounding rock. Originally the crater was a round, almost funnel-shaped depression, 550 feet across and 100 feet deep. During the past 20,000 years, windblown silt and water-laid sediment have gradually filled the depression nearly to the surface. The crater now appears as a shallow, circular pit with a rubbled rim of limestone defining its outline. The uptipped rim, circular shape, and presence of high-pressure minerals, found in a 165-foot hole drilled in the center of the crater serve to confirm its meteoric origin.

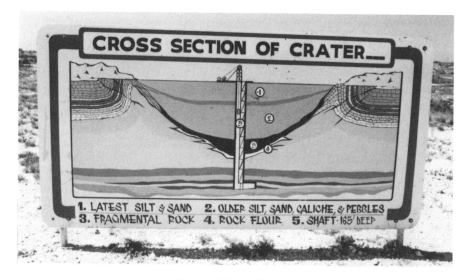

Odessa Meteor Crater.

Odessa Meteor Crater is one of only two well-documented meteor impact craters in Texas, the other being the Sierra Madera Crater located 38 miles northeast of Marathon, Texas on Highway 385.

To get to the Odessa Meteor Crater, follow the signs off Interstate I-20 near Odessa to Meteor Crater Road. At the site are two interpretive plaques, picnic tables, and a walkway with explanatory signs through the crater.

Interstate 40
Oklahoma—Amarillo—New Mexico
182 miles

Interstate 40 bisects the Texas Panhandle from east to west, traverses the High Plains in the center, and rides on deeper-eroded, older rocks at each end near the New Mexico and Oklahoma borders.

From the Oklahoma border to about nine miles west of Shamrock, the highway rides along the drainage topography of the North Fork of the Red River. The river here has carved into colorful red sedimentary rocks of Permian age. Southwest of Shamrock, resistant beds of white anhydrite and gypsum within the Permian sequence are responsible for the low ridges and topographic relief seen here.

The Interstate climbs upward on a gentle slope nine miles west of Shamrock. This incline represents the eastern caprock "escarpment," which in other places farther south is indeed an abrupt cliff edge.

After the climb, the highway is on the High Plains surface, which is the top of the Miocene-Pliocene-aged Ogallala formation—a vast sheet of gravel, sand, and caliche that spread eastward across the

Cross section along Interstate 40 across the northern Panhandle.

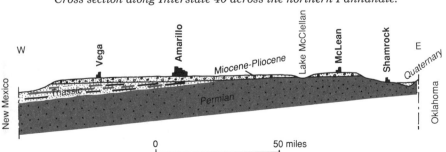

panhandle from the Rockies. This sheet was once a continuous wedge-shaped blanket of sediment that lay at the foot of the Rocky Mountain front from Texas to Montana. This "gangplank" has subsequently been eroded and dissected, but still remains a major geologic feature of North America.

Interstate 40 continues on this flat surface well past Amarillo to within ten miles of the New Mexico border. It is unbroken except for a small erosional depression west of McClean, where the Ogallala has been breached to reveal the underlying Permian rocks around Lake McClellan.

An unusual feature of the High Plains is a vast number of round, flat-bottomed surface ponds, locally called 'playas' or 'playa lakes.' Look for these round, flat, gray, unvegetated, areas where they occur near the highway. An important source of water for early pioneers, these ponds collect rainfall on the flat surface which has few drainages and consequently little run-off. Some water evaporates, the rest settles into the underlying Ogallala sands to form the famous Ogallala aquifer where millions of gallons of fresh water are naturally stored beneath the surface of the High Plains.

From Amarillo, State Highway 136 north leads to Alibates Flint Quarries National Monument and Lake Meredith along the "breaks" of the Canadian River. On Interstate 27–U.S. 87 south of Amarillo is the town of Canyon where the Panhandle-Plains Historical Museum is located. And from Canyon, a short drive eastward leads to Palo Duro Canyon State Park, a geological "must-stop" place, and the crown jewel of the Texas State Park System.

Varicolored sandstone, siltstone, and mudstone beds of Triassic age are encountered along Interstate 40, ten miles from the New Mexico border, where the Canadian River drainage system has eroded through the caprock to expose these older rocks. This area forms the western escarpment of the high plains.

U.S. 84
New Mexico—Farwell—Lubbock—
Post—Snyder—Roscoe—I-20
208 miles

The road from the New Mexico state line southeastward to Lubbock is truly on the FLAT High Plains. Low topography etched into the Ogallala sandstone is seen south of the highway near the town of Lariat where the Blackwater River—an upend portion of the Brazos River drainage system—has carved into the High Plains surface. Low conical hills or knobs southeast of Muleshoe are wind-piled remnant dunes of sand.

Near Littlefield, where the countryside is quite flat, is a sign to a nearby town, well named, Levelland! Though the landscape, as explained in the introduction, has a sound geologic reason for being featureless, it is not a "wasteland," by any means, as the brown to reddish soils produce rich crops of cotton, corn, and wheat. Large grain elevators at most of the small towns along this stretch of road attest to this richness.

The road passes through several oil fields that produce from Paleozoic (Permian) reservoirs in the underlying Permian basin, where thousands of feet of sedimentary rock underlie the High Plains here. Organic material deposited with these sediments is the source of the abundant oil produced today.

From Lubbock southeastward toward Snyder and the U.S. 84 junction with Interstate 20, the highway continues to ride the surface of the High Plains. Watch for surface ponds that dot the landscape. They are somewhat subtle to observe, unless you are actually looking for them. But they do tell the story of how rainfall, geology, flat-topography, slow run-off, and subsurface aquifers all act in concert on the High Plains.

The edge of the plains comes up abruptly on the north side of the town of Post, where U.S. 84 drops off the plains into the erosional topography of the Double Mt. Fork of the Brazos River. Sand and conglomerate of the Miocene-Pliocene Ogallala formation lie directly on colorful, varicolored sand and clay beds of the Triassic Dockum group in nice little roadcuts near Post. A view to the west from the highway shows that the caprock escarpment, though not very tall, is nevertheless vertical and distinct.

The highway continues southeastward for about 20 miles on Triassic varicolored sedimentary rocks that are exposed in the river drainage. The road then climbs onto the High Plains surface again for a few miles on the way to Snyder. And, from Snyder southeastward to I-20 and Roscoe, the road remains on the High Plains surface.

U.S. 287
Amarillo—Childress
117 miles

U.S. Highway 287 transects the High Plains, eastern Caprock escarpment, and the Low Rolling Plains to the east. Between Amarillo and Clarendon the road travels on the High Plains surface, and vistas of flat, immense distances are seen from the car windows. Twenty miles east of the U.S. Highway 287 and Interstate 40 junction is the hamlet of Claude, where the scenic road, Texas Highway 207, heads south toward Silverton. This pleasant sidetrip road crosses a wide expanse of Palo Duro Canyon east of Palo Duro Canyon State Park (see the road description in a later section).

Near Clarendon watch for tan soils, small outcrop exposures of Ogallala sandstone and a rush-and-water-filled surface pond south of the highway. Also, near Clarendon, increased topographic relief is noticeable as the erosive effects of the nearby Salt Fork of the Red River come into play.

Badlands country comes into view 20 miles east of Clarendon where outcrops and roadcuts of tan, Ogallala sandstone are seen. East of Hedley, unmistakable red Permian rocks are encountered as the road now is cutting its way downward through the eastern Caprock escarpment. Note red mesas east of Memphis, and roadcuts south of town in the distinctive red Permian-aged sediments. A bit farther south of Memphis are remnant buttes that stream erosion has left behind—these, again, are of red Permian rocks.

About halfway between Memphis and Childress, Highway 287 crosses the Prairie Dog Town Fork of the Red River (wonderful name!); if water is flowing, it is not hard to understand why early pioneers dubbed this the Red River! The river's sediment load, being composed largely of eroded particles from the surrounding red Permian countryside, impart the River's distinctive color. South of the bridge is a pull-off, which offers an opportunity to look at the sand dunes that have been piled up by the wind. With such an abundant sand source lying about in the frequently dry river bed, it is not surprising the wind would create dunes. All the elements for dune formation are present: wind, a source, and places to trap the sand (vegetation obstacles).

But don't disregard the array of river bars, channels, and gravel piles beautifully exposed in the bed of the (dry) river. This is a real chance to see how the bed load of a stream works and the kinds of deposits flowing water leaves behind. A real lesson in dynamic stream geology can be learned from spending a few minutes looking at the bed of the Prairie Dog Town Fork of the Red River!

Between the river crossing and Childress are continuous red soils, an occasional red outcrop, and low red hills, telling of the presence of Permian-aged sedimentary rocks.

Texas 18
Monahans—Kermit—
New Mexico State Line
32 miles

You will see a lot of sand dunes on this stretch of road, sometimes as far as the eye can see on either side of the highway. Since the Ice Age, the sand has blown eastward out of the Pecos River valley, where abundant sand is available from the river's dry bed load. Notice in many places, particularly north of Kermit, how the orange-tan sand is trapped in piles behind the tough, bushy vegetation. These are called "copse dunes," and illustrate the important role the local plants have in trapping and holding the sand in this area.

The road also passes through several large oil fields. It is particularly noticeable how all the pumps line up in regular order. The spacing of wells is designed to maximize efficiency of production and is dictated by state regulation related to depth, flow rate of oil from the reservoir, and the reservoir's thickness and size.

A few miles north of Monahans look for a shallow quarry on the west side of Highway 18 where brown stream gravels and brown dune sand overlie some rather nice exposures of white, nodular caliche.

View eastward of Caprock escarpment from the picnic area between Silverton and Quitaque. Permian red-bed plains in the distance.

Texas 86, 256, 207
Tulia—Silverton—Claude—
Quitaque—Turkey
51 miles

About 50 miles south of Amarillo, Highway 86 crosses Interstate 27 and U.S. 87 at the town of Tulia. Turn east onto Texas 86, and follow this road as it heads across the High Plains on the way to Quitaque and Caprock Canyons State Park. About 18 miles east of Tulia, Highway 86 passes the head of Rock Creek, seen to the left (north) of the road. This is a site of Ice Age fossil mammals collected by paleontologists from Yale University in 1912. Horses, camels, ground sloths, giant short-faced bear, imperial mammoth, dogs, and prong-horn antelope from this locality all date about seven-hundred-thousand to a million years old.

Texas Highway 207

Four miles west of Silverton is the junction of Texas Highways 86 and 207. Highway 207 is a 48 mile scenic road that heads north to the town of Claude, which is located on U.S. Highway 287. State Highway 207 crosses a wide part of Palo Duro Canyon east of Palo Duro Canyon State Park. Permian red beds and massive gypsum beds can be seen from the highway in the widened canyon. A few miles north of Highway 86, Texas Highway 207 crosses Tule Canyon at Lake MacKenzie, where one can see Ice Age (Pleistocene) Tule formation sediments in roadcuts directly overlying exposures of Triassic Trujillo formation. Look for beautiful, sparkly selenite (gypsum) crystals in the Tule formation in roadcuts north of the Tule Canyon crossing.

The Trujillo formation forms spectacular buttes and small mesas in Tule Canyon. Between Palo Duro and Tule Canyons, Highway 207 crosses the High Plains surface.

Texas Highways 86 - 256

Continuing eastward on Highway 86, the road passes through the town of Silverton, and about 2 miles east of town is the junction of Highways 86 and 256. Here you can make the choice whether to take the longer scenic route (Highway 256), or the shorter, and only a little less scenic route (Highway 86) to Quitaque and Caprock Canyons State Park.

Texas Highway 256

This road heads east and is the more scenic and spectacular descent from the edge of the High Plains, down through the Caprock escarpment to the low rolling Permian plains. The escarpment is encountered nine miles east of Silverton. Watch for the through-park road which turns south off Highway 256 and connects with Highway 86 near Quitaque. This park road runs north-south through Caprock Canyons State Park at the base of the escarpment and provides good continuous views of the cliffs. Note the alluvial fans and talus slopes as you drive along.

Continuation of Texas Highway 86

The road turns southeast from the junction with Texas Highway 256 and heads toward Quitaque. Six miles from the junction is a picnic area that overlooks the colorful edge of the Caprock country and the reddened landscape of the Permian plains to the east. From here you can see canyon and badland exposures of the Ogallala sandstone and caliche forming the top edge of the escarpment. Beneath are the purple-white-gray-tan-red-pink beds of sandstone and shale of the Triassic-age Dockum group of rocks. And, at the base of the cliffs these varicolored Triassic beds give way to bright red sandstones of Permian age that extend eastward for miles.

From the picnic overlook the road then winds its way downward through the whole panorama to give you a spectacular close-up, car-window look at the rocks in nearby hills and roadcuts.

Red-bed Permian geology is seen in hills and soils from the bottom of the escarpment, through Quitaque, and on to the town of Turkey.

But don't miss Caprock Canyons State Park just north of Quitaque. Follow the road signs in town to the park. This is one of the most colorful, geologically interesting and pleasant little parks in the state.

373

View of erosional topography in Caprock Canyons State Park, from Interpretive Center. Color this picture bright red!

Triassic sandstone mesa.

Multiple strings of gypsum amongst red siltstone and mudstone allude to repeated wet-dry cycles in Permian.

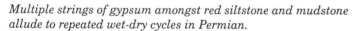

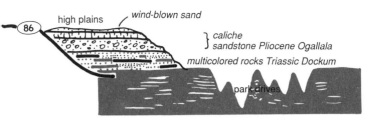

high plains | wind-blown sand

86

} caliche
sandstone Pliocene Ogallala

multicolored rocks Triassic Dockum

park drives

86

256

207

Geologic cross section at Caprock Canyons State Park.

CAPROCK CANYONS STATE PARK

Just a few miles north of the town of Quitaque off U.S. 86 is this bright, colorful park, located along the edge of the Caprock Cliffs. The geology here is similar to that described for Palo Duro Canyon— Caprock Ogallala overlies varicolored Triassic rocks which in turn overlie bright red Permian beds. But the park drive snakes through some of the best close-up outcrops of Permian red beds to be seen in this part of Texas, and the gypsum deposits, all white amongst the red, and tortuously deformed in their softness by extinct geologic forces, are spectacular!

A wonderful, new interpretive exhibit, housed in an open-air rotunda just off the road to the campground is called "250 Million Years at Caprock Canyons." You'll get more geological understanding in 15 minutes at this exhibit than most anywhere else. It covers not only geology, but human history and paleo-biology with excellent drawings and displays.

Contorted white gypsum beds, surrounded by red mudstone in a roadcut on the main park road.

375

Texas 136
Amarillo—Alibates Flint Quarries
National Monument
38 miles

From Amarillo northward, Texas Highway 136 traverses flat to low, barely rolling countryside typical of the High Plains. Dissection of this surface by tributary drainages to the Canadian River becomes more and more apparent as you drive farther north toward the river. At about 36 miles north of Amarillo (from the intersection of Amarillo Boulevard [U.S. 66] and the Fritch Highway [Texas 136]), watch for the turnoff to Alibates Flint Quarries National Monument, Lake Meredith Recreation Area and Bates Canyon. Turn left (west) and follow the signs to the monument. These areas can be visited anytime, but the monument and its Indian Flint Quarries are only open via park service guided tours.

As you drive into the monument area, red Permian beds are exposed by the erosive action of the Canadian River, which has cut an impressive path across the Texas Panhandle, completely dissecting the High Plains. This segment of colorful erosional topography, canyon country, and river morphology is called "The Breaks" by local folks. The High Plains are so vast that a "break" in the surface is noteworthy!

Though headward erosion by streams has incised the High Plains and moved the eastern escarpment to the west many miles in the last few million years, the valley of the Canadian River is apparently not the single product of such headward erosion. Certainly some headward erosion by the Canadian River accounts for the river's incision through the Ogallala formation and the High Plains surface. However, recent geological research indicates the Canadian River valley is located in a long, low trough which is the result of subsidence from the dissolution of bedded salt layers in underlying Permian strata. As ground water reached the salt layers buried 2,000 feet beneath the surface the salt was dissolved, and the rocks above collapsed into the void, resulting in a corresponding surface trough. Surface drainage was naturally drawn to this low area, creating the location for the subsequent Canadian River. The processes of dissolution, salt removal and continued subsidence are still active, judging from the high salinity (3,000 parts per million) of Canadian River waters entering Lake Meredith. Sinkholes have appeared historically along the Canadian River valley, confirming that salt dissolution continues to be an active local process accompanied by subsidence.

ALIBATES FLINT QUARRIES
NATIONAL MONUMENT

At the north end of the Panhandle, tucked amongst the colorful eroded edges of the High Plains carved by the Canadian River, is one of those little-visited, but delightful small national monuments— Alibates Flint Quarries, the only national monument in Texas. The monument is named from Alibates Creek, in turn after cowboy Allen "Allie" Bates, who lived in a line camp at the quarry site in the late 1800s.

Geologic section at Alibates National Monument.

This historical monument, only 36 miles north of Amarillo, a few miles off Highway 136, preserves 550 flint quarries and Indian ruins, where for 12,000 years, until about 1870, Indians dug out multi-colored flint to make arrowheads, knives, hammers, and awls for their own use and to trade. The Alibates flint, banded and mottled red, pink, pale blue, pale purple, gray, brown, white, and black, was highly prized and widely traded all across the Indian trade network of North America. Alibates flint has been found in archeological sites as far away as Montana, the Great Lakes, and all across Texas and the Great Plains.

Geologically, the Alibates flint comes from ledges of the Alibates dolomite, which is part of the Permian-age Quartermaster formation. The dolomite has been partly replaced by silica solutions to form chert, a fine-grained variety of quartz. "Flint" is a term applied to dark chert, or highly-colored chert.

The bright red siltstone, claystone, sandstone, and white cherty dolomite rocks seen everywhere in the monument from the shoreline level of Lake Meredith up through the dissected hills are the Permian-aged (about 225 million years) Quartermaster formation. At the tops of the hills and along the road driving into the monument are white sandstone and caliche ledges and blocks of the Ogallala formation of Miocene-Pliocene age — about four to ten million years old. At the monument, conical, red haystack-shaped hills are commonly topped

377

by a protective cap of Ogallala sandstone; in colorful contrast, white sandstone blocks rain down the sides of the red hills.

To reach the Monument, watch for the directional sign on Highway 136, at 36 miles north of Amarillo, or six miles south of the town of Fritch. Turn west, and go about three to four miles to a fork in the road. The left fork goes to McBride Canyon Recreation site on Lake Meredith. Take the right fork to Alibates Monument. Though you can drive to the monument and look at the scenery, the quarries and ruins can be reached only on ranger-guided tours by reservation. Write to Lake Meredith National Recreational Area, Box 1438, Fritch, Texas 79036, or call (806) 857-3152 or (806) 865-3874.

Texas 217
Canyon—Palo Duro
Canyon State Park
18 miles

The road begins in the town of Canyon, which is located one mile west of Interstate 27, in a Pleistocene basin. The town is appropriately named, as it rests at the headwaters of the canyon cut by the Prairie Dog Town Fork of the Red River, which runs through town in its eastward flow to Palo Duro Canyon. But before driving to Palo Duro Canyon, do not miss stopping at the Panhandle-Plains Historical Museum, located only a few blocks west of Interstate 27. The geologic exhibits are superb and cover geologic time, fossils, the Amarillo Mountains, the Ogallala aquifer, High Plains ponds, Palo Duro Canyon, and oil and gas, in addition to colorful and up-to-date exhibits on history, Indians, firearms, etc. This is an excellent museum, and is worth the stop.

East of Canyon and Interstate 27, Highway 217 climbs out of the town's basin and heads across the High Plains, where waving grass and crops are seen for miles on either side of the road. It is a familiar scene in this country, and the vastness and flatness are in sharp contrast for what is encountered ahead. For suddenly, the High Plains are abruptly terminated in colorful cliff walls that drop 700 - 800 feet. Pass through the entrance gate at Palo Duro Canyon State Park, but you will still not be prepared for that first view of the scene ahead.

Caprock edge of Timbercreek Canyon at Park entrance. High Plains surface to the right.

PALO DURO CANYON STATE PARK

Palo Duro Canyon sneaks up on you! As you drive eastward from Interstate 27, the High Plains spread to the horizon on either side of Highway 217. And, just before entering the Park, some hint of what is to come is seen in a small side canyon, Timbercreek Canyon, to the right (south) of the highway where ledges of caliche lie over steep-walled cliffs of Ogallala sandstone. But, it is not until you have paid your entrance fee and driven to the overlook that Palo Duro Canyon's size, depth, and color suddenly appear before you.

Palo Duro means "hard wood" in Spanish, named for the canyon's junipers from which Indians made their "hardwood" bows.

Before becoming a state park, Palo Duro Canyon had a rich history, including 12,000 years of Indian occupation. The Spanish explorer Coronado saw the canyon in 1541, and while leading a U.S. army expedition, Captain Randolph Marcy wrote admiringly of it in 1852. The Battle of Palo Duro Canyon in 1874 pitted Colonel Ranald Mackenzie's troops against Comanche warriors, ending the so-called Red River War. Only two years later, Charles Goodnight, famous for spearheading Texas–Wyoming cattle drives, began his JA Ranch operation in the canyon. Panhandle citizens' interest in Palo Duro Canyon as a recreation and picnic area led to a failed attempt to make it a National Park, but succeeded in the creation of the crown jewel of the Texas State Park System in 1934.

But, to know Palo Duro Canyon is to know its geology. The closeness of the raw rock exposures on the steep descent road to the canyon floor almost force you to think about geology. Actually, two main geological stories are written in Palo Duro's colorful walls and 800 foot depth. The first is a tale of dynamic processes—floods, streams, erosion, down-cutting—those forces that cut the canyon itself out of the body of the High Plains. The second story is of ancient environments, sands, and sediments, and long-extinct animals, written in the rocky pages of stratigraphy exposed on the canyon walls.

Before descending into the canyon, stop at the Visitor's Interpretive Center located on the Canyon Rim near the first overlook turnoff. A review of the geology and natural history exhibits here nicely sets the stage for your trip through the Park.

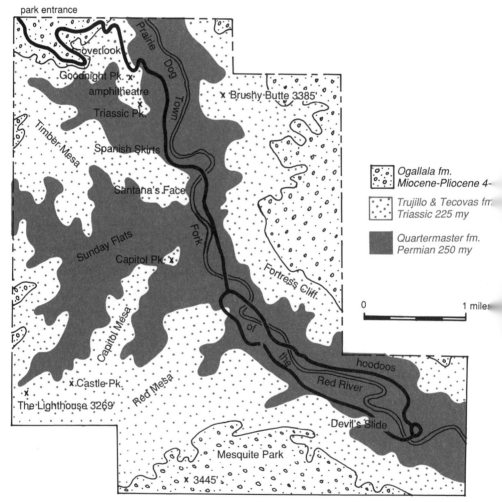

Geologic map of the northern part of Palo Duro Canyon State Park.

Canyon Cutting

Rivers cut canyons. But looking at the Prairie Dog Town Fork of the Red River, it is hard to believe this languid little stream carved out Palo Duro Canyon. However, it did—in two ways.

First, about ten million years ago the Rockies were given an extra shove upward, renewed erosion occurred, and deposits of sand, silt, and pebbles were spread eastward across the Panhandle. The Ogallala formation was the product. Then, in the Ice Age of the Pleistocene, about a million years ago, sea level sank because much world-wide water was held in continental glaciers. It was also wetter and rainier. Rivers not only carried a lot more water, but their mouths were now several hundred miles farther "offshore." Rivers try to balance their profile depending upon their water load, sediment load, and distance they have to travel. Change any one of these factors, and rivers adjust the others to "balance." Pleistocene rivers adjusted to high water load and longer distance to travel by changing their profile by down-cutting in their upper reaches. So little old Prairie Dog Town Fork of the Red River was a lot bigger then than now, and it had an erosion job to do!

Second, the creek is still eroding the canyon bit by bit. You have to be here after a High Plains downpour to see the creek swell, listen to the boulders thunk together, to understand the power of small streams in flash flood. Over geologic time, little bits of erosion can add up to a lot — even a Palo Duro Canyon!

Other examples of erosion in the canyon are abundantly evident. Note the rills and riffles in the red, Quartermaster formation portion of the Spanish skirts, while just above, in the Tecovas, the slopes are rather smooth. This contrast is the result of differential erosion — rocks don't all respond to erosion in the same way. Note how the Trujillo sandstone, a hard rock — harder than surrounding mud-stones — forms cliffs and ledges on the Lighthouse, Santana's Face, and many mesa edges throughout the Park. It simply does not erode as fast as the mudstones.

The Rock Garden at the south end of the park drive is a jumbled pile of Trujillo sandstone blocks that are the resistant leftovers after the mudstone beneath them has been eroded away.

Differential erosion also explains the wonderful pedestals or "hoodoos" seen in many places in the canyon. A hard cap of sandstone "protects" the pedestal underneath from eroding as fast as nearby unprotected mudstone. Many sandstone slabs are derived from higher up in the cliff walls and have slid downward over the mudstone which they now protect from erosion.

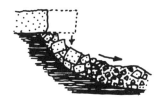

Another erosion process, the local downslope movement of rocks and sediment (mass wasting), creates piles of debris known as talus slopes at the bases of hills, a common feature in the canyon. Rocks fall down fast as well as creep down slowly under the influence of gravity.

The Story in the Canyon Walls

The stratified, layered, sedimentary rocks exposed in the walls of Palo Duro Canyon are filled with information that tell a story of conditions on earth millions of years ago when these rocks were first laid down as fresh sediment.

The packages of distinctive color bands are easily observed in the canyon walls, and are known as formations by geologists, because

Cross section of strata in Palo Duro Canyon.

Miocene-
Pliocene — Ogallala — white caliche
4-10 my — Trujillo — tan conglomerate, sandstone
gray, pink, white shale, siltstone, sandstone
dark gray sandstone

Triassic — Tecovas — varicolored(purple, yellow, white, gray, pink, tan, brc
225 my — shale, siltstone, sandstone

Permian — Quartermaster — bright red sandstone, shale
250 my — white gypsum
Prairie Dog Town

View to southeast of Palo Duro Canyon from scenic overlook and Visitors' Center near Park entrance.

each package can be recognized and mapped as distinctive units of time and deposition.

The bright red rocks at the floor and on the lower slopes of the canyon are the oldest rocks exposed, laid down in shallow marine waters and drying flats at the edge of the seas in Permian time, 245 - 286 million years ago. The wave-formed ripple marks and distinctive white gypsum (calcium sulfate—an evaporative mineral) indicate shallow sea and drying-flat conditions, respectively. The types of cross-bedding in certain sandstones also reaffirm a shallow sea interpretation for the environment in which these rocks were deposited. Red Quartermaster rocks form the Spanish skirts, seen from the overlook near the park entrance.

Capitol Peak. The upper and middle section is Triassic Tecovas formation. Lower red slopes are Permian Quartermaster formation.

Fold in red Permian Quartermaster formation.

Above the Quartermaster red-beds is a varicolored rock package, consisting mostly of slope-forming shales, but also of ledges of sandstone and siltstone. Colors of purple, yellow, white, gray, pink, tan, and brown flick on and off in varied light conditions of the passing day. These rocks comprise the middle slopes of most mesas and hills in Palo Duro, and are of Triassic age, 208–245 millions years ago. These are known generally in the Panhandle area as Dockum group rocks, but are broken into two formations in the canyon, the Tecovas and the Trujillo. Tecovas shales are the lower, colorful slopes just above the Permian red beds, whereas the Trujillo is easily recognized by the ledge-forming sandstone that tops many of the canyon's middle-level mesas.

Pile of large sandstone blocks at end of Park Road, called appropriately "Fallen City." The boulders are Trujillo sandstone derived from the upper cliffs.

Mudstones in the Tecovas were laid down in streams and swamps, judging from the fossils of crocodile-like reptiles (phytosaurs), primitive amphibians and fish. The multi-hued colors indicate oxidation and drying and wetting cycles so typical of the varied conditions found in streams and their surrounding swamps. The Tecovas can be seen in the upper part of the Spanish Skirts, on Capitol Peak, and The Devil's Slide.

The Trujillo is easily picked out by the conspicuous sandstone ledge that rests on top of many mesas. Coarse sand and river-type cross-bedding in the Trujillo sandstone tell us it was deposited in an ancient stream bed. Fossils of primitive amphibians, wood, reptiles, and leaves are found, but are not common in the Trujillo. Santana's Face, the cap on the Lighthouse, the Rock Garden, and numerous pedestal caps are all Trujillo sandstone.

At the top of the Canyon is a cliff-forming ledge of tan sandstone, opal-cemented siltstone, conglomerate, and at the very top, a white caliche ledge. This is the Ogallala formation of Miocene-Pliocene age of 4 to 10 million years ago. The caliche formed toward the end of a long period of progressive drying, and environmental conditions in the Panhandle changed from a savannah about ten million years ago to a steppe by about four million years ago. You will note a lot of time is missing between the Trujillo and Ogallala. Either the rocks representing about 200 million years of time were eroded away, or they were never deposited; whatever the case, a great unconformity is represented by the mere line between the multi-colored upper beds of the Trujillo and the lower tan beds of the Ogallala. The missing time represents most of the great Age of Dinosaurs and most of the great Age of Mammals.

As discussed previously, the Ogallala was laid down as a wedge of sediment in front of the Rockies. The caliche layer at the top represents alternating drying and wetting conditions in the soil near the surface, so typical of desert conditions in the High Plains, even today. The sandstones in the Ogallala are vital to the entire High Plains as a water aquifer.

Fossils of saber-tooth cat, bone-crushing dogs, early mastodons, rhinoceroses, horses, long-necked camels, and tortoises have been preserved in Ogallala strata.

You are standing on Ogallala rocks, mostly caliche, at the overlook, and you drive down through them on the road to the canyon floor. Fortress Cliff that forms the east canyon rim has an impressive wall of Ogallala formation at its top edge.

For more information about the geology and history of Palo Duro Canyon, look for these pamphlets for sale at the Goodnight Trading Post:

1. *The Geologic Story of Palo Duro Canyon* by William H. Matthews, III, 1983 edition, Guidebook 8, Bureau Economic Geology, University of Texas, Austin, 51 pp.
   In-depth presentation for those interested in stratigraphy, paleontology, minerals, local features, maps, photos, illustrations.

2. *Palo Duro Canyon State Park* by R. Franks and G. Zulauf, 1966, 45 pp.
   General guide, includes some geology, more history, things to do, maps and trails.

3. *The Story of Palo Duro Canyon,* Panhandle-Plains Historical Review, 1978, vol. 51, 243 pp. plus maps and plates.
   A collection of articles on the geology, paleontology, archeology, vegetation and history of the Canyon.

Texas 349, U.S. 87, Interstate 27
Midland—Lamesa—Lubbock—
Plainview—Canyon—Amarillo
250 miles

This is a true High Plains route, as the 250 mile stretch between Midland and Amarillo is entirely on the flat surface of the Llano Estacado, the staked plains. Undulations in this surface are mainly due to eastward flowing stream courses that traverse the High Plains — these are headwaters for several of Texas' major rivers such as the Brazos, Colorado, and Red. The north-south highway crosses these drainages at nearly right angles, as they flow eastward off the High Plains, to join other streams to become the main rivers that eventually empty far southward into the Gulf of Mexico.

Many surface ponds, a unique feature of the High Plains, are seen along the length of Texas 349-U.S. 87 between Midland and Amarillo. The unvegetated, gray, round, flat areas fill with water after thunderstorms, only to dry up again as the ponds evaporate and water sinks into the ground to recharge the underlying Ogallala aquifer.

About the only rocks to break the flat surface are several low, isolated mesas to the west and northeast of the town of Tahoka. These 50-foot-high mesas are remnants of hardened, sandy, Cretaceous shoreline deposits that, because of their hardness, survived erosion and now stand above the surrounding younger Ogallala sand and caliche deposits.

A marked change in vegetation is apparent from south to north on this road. The sparse, dry, mesquite grazing country around Midland contrasts sharply with the lush grass and cropland farther north. The average annual precipitation is not that much different between Amarillo and Midland, but the northern Panhandle receives much of its precipitation in the winter in the form of snow, whose water slowly saturates the underlying soil and is retained to provide moisture for plant growth. But drying winds and low overall rainfall still make trees rare on the High Plains, except in the low areas along stream drainages.

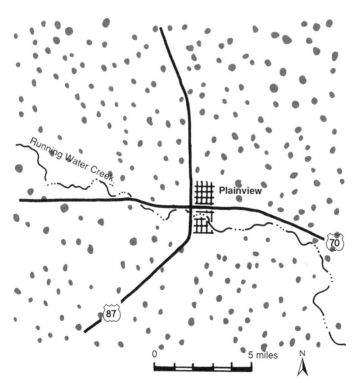

Map shows profusion of surface ponds on High Plains around Plainview, Texas.

Maps and Information

Geological maps of Texas are available from several sources. If interested, you may want to write to the following organizations for a list of available maps and prices.

United States Geologic Survey (USGS)
Western Distribution Branch
Box 25286 Federal Center Building 41
Denver, Colorado 80225

Publications Office
Bureau of Economic Geology (TBEG)
The University of Texas
Box X
University Station
Austin, Texas 78713

AAPG Bookstore
The American Association of Petroleum Geologists
Post Office Box 979
Tulsa, Oklahoma 74101

Texas Mosaics (TM)
Box 5273
Austin, Texas 78763

Dino Productions (DP)
Post Office Box 3004
Englewood, Colorado 80155

A few key maps are:

Texas Mineral Resources 1:1,000,000 - 1979 (TBEG)

Texas Topographic Map 1:1,000,000 - 1965 (USGS)

Geologic Atlas of Texas - 32 maps 1:250,000 (TBEG)

Texas Energy Resources 1:1,000,000 - 1976 (TBEG)

Texas Geothermal Resources 1:1,000,000 - 1982 (TBEG)

Texas, Big Bend National Park 1:1,000,000 - 1971 (USGS)

Geologic Map of Big Bend National Park, Texas 1:62,500 - 1967 (TBEG)

Geologic Highway Map of Texas 1:2,000,000 - 1973 (AAPG)

The Texas Landscape - Geographic Provinces of Texas 1:2,000,000 - 1986 (TM)

The Basins and Ranges - The Mountains of Far Western Texas 1:500,000 - 1986 (TM)

Pathway to the Dinosaurs - Map #3 Texas - 1989 (DP)

Land Resources of Texas - 1977 (TBEG)

There are many detailed guidebooks as well as a plethora of books and papers on Texas geology. These publications may be purchased from the following organizations. Lists and prices are available upon request:

Publications Office
Bureau of Economic Geology
The University of Texas
University Station, Box X
Austin, Texas 78713

AAPG Bookstore
American Association of
Petroleum Geologists
Post Office Box 979
Tulsa, Oklahoma 74101

United States Geological Survey
Western Distribution Branch
Box 25286
Federal Center Building 41
Denver, Colorado 80225

The Geological Society of America
3300 Penrose Place
Post Office Box 9140
Boulder, Colorado 80301

Regional geologic guidebooks are generally available from the following local geological societies. Write for lists and prices for your area of interest.

Abilene Geological Society
Post Office Box 974
Abilene, Texas 79604

Graham Geological Society
Post Office Box 862
Graham, Texas 76046

Austin Geological Society
Post Office Box 1302
Austin, Texas 78767

Houston Geological Society
7171 Harwin Drive, Suite 314
Houston, Texas 77036

Corpus Christi Geological Society
Post Office Box 1068
Corpus Christi, Texas 78403

North Texas Geological Society
Post Office Box 1671
Wichita Falls, Texas 76307

Dallas Geological Society
Suite 170,
One Energy Square
Dallas, Texas 75206

Panhandle Geological Society
Post Office Box 2473
Amarillo, Texas 79105

East Texas Geological Society
Post Office Box 216
Tyler, Texas 75710

San Angelo Geological Society
Post Office Box 2568
San Angelo, Texas 76902-2568

El Paso Geological Society
c/o Department of
Geological Sciences
University of Texas at El Paso
El Paso, Texas 79968

South Texas Geological Society
900 NE Loop 410, Suite D-100
San Antonio, Texas 78209

West Texas Geological Society
Post Office Box 1595
Midland, Texas 79702

Credits

All photographs were take by the author, and all maps and illustrations were drawn by the author specifically for this book.

Many illustrations are the author's own derivation, some are modified from other geologists' published books and papers, which are cited below, and used with kind permission of authors/publishers.

Geologic maps were redrawn from the 1:250,000 scale sheets of the Geologic Atlas of Texas, published by the Bureau of Economic Geology, the University of Texas, Austin, Texas or the Geologic Highway Map of Texas, published by the American Association of Petroleum Geologists, Tulsa, Oklahoma.

Text material has been gleaned from the abundant literature of Texas geology. Much of it comes from guidebooks, pamphlets, and papers published by the Bureau of Economic Geology, which are listed in the section entitled Where to Get Information. The format of this book is not conducive to citing everyone of these literature sources, but the author gratefully acknowledges and recognizes the many men and women who have devoted careers to unraveling the geology of this state and who have published their observations and thoughts.

Introduction (The Big Picture)

Geologic Time Scale, page xiv
Ages from 1983 Geologic Time Scale, Geologic Society of America, Boulder Colorado.

Maps in the March Through Geologic Time, pages 3-10
Sketches based on maps in Cook, T.D., and Bally, A.W., Stratigraphic Atlas of North and Central America, 272 p., copyright 1975 by Shell Oil Co., Princeton University Press, N.J., maps used by permission of Princeton University Press.

Geologic Map of Texas, page 12
Compiled from Geologic Highway Map page of Texas, Map #7, American Association Petroleum Geologists, Tulsa, Ok, and Geologic Atlas of Texas, 38 map sheets, 1:250,000, Univ. Texas, Austin, Bureau Economic Geology.

Maps - Continents Through Time, page 14
Compiled from Dietz, R.S., and Holden, J.C., 1970, Journal Geophysical Research, v. 75, no. 26; and Bambach, R.K., Scotese, C.R., and Ziegler, A.M., 1980, Amer. Scientist, Jan. issue.

Cretaceous Seaway Map, page 17
Compiled from maps in Caldwell, W.G.E., 1975, The Cretaceous System in the Western Interior of North America, the Geological Association Canada, Special Paper #13.

Physiographic Map of Texas, page 22
Modified from Kier, R.S., Garner, L.E., Brown, L.F., Jr., 1977, Land Resources of Texas, Special Report, Univ. Texas, Bur. Econ. Geology.

Average Rainfall Map, page 23
Adapted from Kier, et al, 1977.

Vegetation Map of Texas, page 24
Compiled from Texas Almanac 1984 - 1985, The Dallas Morning News, Dallas, Texas; Kier, et al., 1977; and leaflet 492, Texas A&M Exper. Stat. Extension Serv. and U.S. Dept. Agriculture.

Map - Structural Geology of Texas, page 27
Modified from Kier, et al., 1977.

Map of Oil and Gas Reservoirs, page 28
From Tyler, N., et al, 1984, Oil Accumulation, Production Characteristics and Targets for Additional Recovery in Major Oil Reservoirs of Texas, Univ. Texas, Austin, Bur. Econ. Geology, Geol. Circ. #84-2.

Oil and Gas Map of Texas, page 29
Adapted from Vlissides, S.D., 1964, Map of Texas showing Oil and Gas Fields, Pipelines and Areas of Exposed Basement Rocks, U.S. Geological Survey, Map OM-214.

Charts of Texas Crude, and U.S. Demand, Supply, Imports, page 30
Compiled from various sources, including Amer. Petroleum Inst., Amer. Assoc. Petroleum Geologists and Amer. Geological Inst. data.

Energy Resources Map, page 32
Based on Kier, et al, 1977.

Mineral Resources Map, page 33
Based on Kier, et al., 1977.

Gulf Coast

Cross Sections of History and Development of the Gulf of Mexico, page 37
From Halbouty, M.T., 1979, Salt Domes, Gulf Region United States and Mexico, 2nd ed., Gulf Publ. Co., Houston, Texas.

Aerial View of Meander Belt, page 40
Drawn from an air photo in Bernard, H.A., et al, 1970, Recent Sediments of Southeast Texas, Guidebook 11, Univ. Texas, Austin, Bur. Econ. Geology.

Map of Shoreline Positions since Cretaceous, page 42
From Walker, H.J., and Coleman J.M., 1987, Atlantic and Gulf Coastal Province, In Graf. W.L., ed., Geomorphic Systems of North America, Geological Soc. Amer., Centennial Spec. Vol. 2, p. 65.

Maps of Geologic History of Texas Barrier Islands, pages 43, 44
From Weise, B.R., and White, W.A., 1980, Padre Island National Seashore, Guidebook 17, Univ. Texas, Austin, Bur. Econ. Geology.

Map and Cross Section of Barrier Islands along the Texas Coast, page 45
From Scott, A.J., et al., 1984, Modern Depositional Systems, Texas Coastal Plain, Ann. Guidebook, Amer. Assoc. of Petroleum Geologists, Tulsa, Ok.

Other Hurricane Data, page 47
From Nummedal, D., 1982, Hurricane Landfalls along the Northwest Gulf Coast, In Sedimentary Processes and Environments along the Louisianna - Texas Coast, Field Trip Guidebook, Geological Soc. Amer. Ann. Meeting, New Orleans, LA.

Hurricane Map, page 48
Courtesy of Barry Willey, Dept. of the Army, U.S. Army Corps of Engineers, Galveston, Texas.

Salt Dome and Salt Maps, pages 50, 51, 52
Adapted from Halbouty, 1979.

Southeast Texas

Geologic map of Southeast Texas, page 54
From Geologic Highway Map of Texas, Map #7, Amer. Assoc. Petroleum Geologists, Tulsa, Ok.

Geologic Map of Trinity River, page 59
From Geologic Atlas of Texas, Houston Sheet, 1:250,000, Univ. Texas, Austin, Bur. Econ. Geology.

Geological Map of Austin, page 62
From Trippet, A.R., and Garner, L.E., 1976, Guide to Points of Geologic Interest in Austin, Guidebook 16, Univ. Texas, Austin, Bur. Econ. Geology.

Cross Section of Austin, page 64
From Trippet and Garner, 1976.

Barton Springs Cross Section, page 65
From Trippet and Garner, 1976.

Pilot Knob Cross Section, page 66
From Trippet and Garner, 1976.

Map of Subsurface Faults in Houston Area, page 67
From Etter, E.M., ed., 1981, Houston Area Environmental Geology, Guidebook, Houston Geol. Soc., Houston, Texas.

Map of Subsidence in Houston-Galveston Area, page 68
From Fisher, et al., 1972, Environmental Geologic Atlas of the Texas Coastal Zone. Galveston-Houston Area; Univ. Texas, Austin, Bur. Econ. Geology.

Houston-Galveston Environmental Map, page 71
From Fisher, W.L., et al., 1972.

Map of Water Circulation in Galveston Bay, page 72
From Fisher, et al., 1972.

Map and Cross Section of Galveston Island Environments, page 76
From Fisher, et al., 1972.

Damon Mound Maps, page 83
From Etter, E.M., ed. 1979, Damon Mound Field Trip, Guidebook, Houston Geological Soc., Houston Texas.

Southwest Texas

Geologic Map of Southwest Texas, page 84
From Geologic Highway Map of Texas, Map #7, Amer. Assoc. Petroleum Geologists, Tulsa, Ok.

Geologic Map of San Antonio, page 86
From Geologic Atlas of Texas, San Antonio Sheet, 1:250,000, Univ. Texas, Austin, Bur. Econ. Geology.

Cross Section of San Antonio Groundwater, page 87
From Ewing, T.E., ed., 1989, Urban Geology of San Antonio Area, Texas, Guidebook, South Texas Geological Soc., San Antonio, Texas.

Padre Island National Seashore, page 98
Reference: Weise, B.R., and White, W.A., 1980, Padre Island National Seashore, Guidebook 17, Univ. Texas, Austin, Bur. Econ. Geology.

Map of Aransas National Wildlife Refuge Area, page 102
Compiled from Geologic Atlas of Texas, Corpus Christi Sheet, and Beeville - Bay City Sheet, 1:250,000, Univ. Texas, Austin, Bur. Econ. Geology.

Map of Port Isabel - South Padre Island, page 108
From Geologic Atlas of Texas, McAllen - Brownsville Sheet, 1:250,000, Univ. Texas, Austin, Bur. Econ. Geology.

Central Texas

Geologic Map of Central Texas, page 112
From Geologic Highway Map of Texas, Map #7, Amer. Assoc. Petroleum Geologists, Tulsa, Oklahoma.

Cross Section - Geologic Story of Llano Region, page 123
From Peterson, J.F., 1988, Enchanted Rock State Natural Area, A Guidebook to the Landforms; Terra Cognita Press, San Antonio, Texas.

Map of Precambrian Geology of Llano Uplift, page 130
From West Texas Geological Soc., 1980, Geology of the Llano Region of Central Texas, Guidebook #80-73.

Map of Volcanic Bodies in Uvalde Area, page 136
Compiled from Geologic Atlas of Texas, San Antonio and Del Rio Sheets, 1:250,000, Univ. Texas, Austin, Bur. Econ. Geology.

Geologic Map of the Del Rio Area, page 138
From Geologic Atlas of Texas, Del Rio Sheet, 1:250,000, Univ. Texas, Austin, Bur. Econ. Geology.

Geologic Map of Marble Falls - Johnson City and Pedernales Falls State Park, page 147
From Geologic Atlas of Texas, Llano Sheet, 1:250,000, Univ. Texas, Austin, Bur. Econ. Geology.

Geologic Map of Marble Falls Area, page 158
From Geologic Atlas of Texas, Llano Sheet, 1:250,000, Univ. Texas, Austin, Bur. Econ. Geology.

Geologic Map of Enchanted Rock Batholith, page 164
From Geologic Atlas of Texas, Llano Sheet, 1:250,000, Univ. Texas, Austin, Bur. Econ. Geology.

Fracture Zones - Enchanted Rock, page 166
From Peterson, 1988.

Map of Backbone Ridge, page 170
From Barnes, V.E., at al., 1972, Geology of the Llano Region and Austin Area, Guidebook 13, Univ. Texas, Austin, Bur. Econ. Geology.

Cross Section of Backbone Ridge, page 171
From Matthews, W.H., III, 1963, The Geologic Story of Longhorn Cavern, Guidebook 4, Univ. Texas, Austin, Bur. Econ. Geology.

Geologic Map of Hill Country Drives, page 180
Compiled from Geologic Atlas of Texas, Llano Sheet and San Antonio Sheet, 1:250,000, Univ. Texas, Austin, Bur. Econ. Geology.

Northeast Texas

Geologic Map of Northeast Texas, page 184
From Geologic Highway Map of Texas, Map #7, Amer. Assoc. Petroleum Geologists, Tulsa, Ok.

Geologic Map of Dallas - Ft. Worth, page 188
From Geologic Atlas of Texas, Dallas Sheet, 1:250,000, Univ. Texas, Austin, Bur. Econ. Geology.

Northcentral Texas

Geologic Map of Northcentral Texas, page 210
From Geologic Highway Map of Texas, Map #7, Amer. Assoc. Petroleum Geologists, Tulsa, Oklahoma.

Geologic Map of Weatherford - Possum Kingdom Lake Area, page 225
From Geologic Atlas of Texas, Dallas Sheet and Abilene Sheet, 1:250,000, Univ. Texas, Austin, Bur. Econ. Geology.

Map, Block Diagram and Cross Sections: Environments during Pennsylvanian, pages 227, 228, 229
From Brown, L.F., Jr., and Wermund, E.G., eds, 1969, Guidebook to Late Pennsylvanian Sediments in Northcentral Texas, Dallas Geological Soc., Dallas, Texas.

Geologic Map of Quanah - Copper Breaks State Park, page 239
From Geologic Atlas of Texas, Wichita Falls - Lawton Sheet, 1:250,000, Univ. Texas, Austin, Bur. Econ. Geology.

Geologic Map of Dinosaur Valley State Park Area, page 242
From Geologic Atlas of Texas, Dallas Sheet, 1:250,000, Univ. Texas, Austin, Bur. Econ. Geology.

West Texas

Geologic Map of West Texas, page 254
From Geologic Highway Map of Texas, Map #7, Amer. Assoc. Petroleum Geologists, Tulsa, Ok.

Maps of the Mountains of West Texas, page 256
Adapted from Official Texas Highway Travel Map, 1988, Texas State Dept. Highways and Public Transportation, Austin, Texas.

Cross Section of Balmorhea State Park, page 261
From West Texas Geological Soc. Guidebook, 1985, Structure and Tectonics of Trans - Pecos Texas; from original drawing by Bruce Pearson.

Cross Section of Franklin Mountains, page 264
Based on Richardson, G.B., 1909, Description of the El Paso District, Texas, U.S. Geological Survey Geologic Atlas, Folio 166.

Geologic Map along U.S. 62-180, pages 268-269
From Geologic Atlas of Texas, Van Horn-El Paso Sheet, 1:250,000, Univ. Texas, Austin, Bur. Econ. Geology.

Sketch Map - El Paso to Guadalupe Mountains National Park, page 270
From West Texas Geological Soc., 1968, Delaware Basin Exploration, Guidebook #68-55.

Map of Paisano Volcano, page 273
From Walton, A.W., and Henry, C.D., eds., 1979, Cenozoic Geology of the Trans - Pecos Volcanic Field of Texas, Guidebook #19, Univ. Texas, Austin, Bur. Econ. Geology.

Map and Cross Section of Marathon Uplift, pages 285, 286
From McBride, E.F., 1988, Geology of the Marathon Uplift, In Geological Soc. Amer. Centennial Field Guide, South-Central Section.

Map of Roads in Big Bend National Park, page 292
Based on brochure by Pearson, J., 1980, Road Guide to Paved and Improved Dirt Roads of Big Bend National Park; Big Bend Natural History Assoc. and National Park Serv., Paragon Press.

Geologic Map Big Bend National Park, page 293
From Geologic Map of Big Bend National Park; Brewster County, Texas, 1966, Univ. Texas, Austin, Bur. Econ. Geology.

Glossary

accretion: addition of material; as filling of a stream bed, or growth of a continent by addition of sediment over time.

algal mat: in tidal flats where algae grows in spaces between sand grains, binding the grains together to form a tough mat.

algal mound: thickening in a limestone layer due to local growth of algae.

alkali, alkali flat, alkali lake: in desert areas where a flat lake or plain is encrusted with salts - sodium and potassium carbonate, sodium chloride, that are concentrated by evaporation and poor drainage.

alluvial fan: fan-shaped deposit of stream-deposited sand and gravel found at the base of steep slopes and mountain fronts.

alluvial outwash: the material 'washed-out' from a mountain range, by streams.

alluvial plain: extensive flat or gently sloping surface of loose sand and gravel, deposited by streams.

ammonite: large mollusk that looks like a snail coiled in one plane.

amphibian: cold-blooded animals that breathe by gills early in life, and later by lungs. For example, frogs.

anhydrite: an evaporation mineral, calcium sulfate. It is usually white, and is related to gypsum.

anthracite: hardest coal, has a semi-metallic luster, and is 97-98% carbon.

anticline: arched-up fold of rocks.

aquifer: porous and permeable rock that stores and conducts groundwater in amounts useable by people.

ash-flow tuff: deposit of hot, loose ash in a gaseous cloud that flows rapidly from a volcano.

asteroid belt: a zone of small bodies that orbit the sun mostly between Mars and Jupiter.

atmospheric pressure: the pressure exerted by the column of air over a point on the earth's surface.

badlands: intricately carved landscape dissected by streams.

barchan dune: crescent-shaped sand dune which forms on hard, flat surface where sand supply is limited.

Barrier Island; long, narrow coastal island, separated from the mainland by a shallow water lagoon.

basalt: dark-colored, fine-grained igneous rock, rich in iron, calcium, and magnesium; it commonly occurs as lava flows from volcanoes.

basin: low area in earth's crust where sediments accumulate; locally, a low area between mountain ranges.

batholith: a large body of igneous rock intruded into overlying rock.

bay: curved, open indentation of the coast, commonly where river mouths have been flooded by rising sea level.

bayou: a creek or small stream.

bed, bedded: smallest layer of sedimentary rock, usually a layer from one depositional event, such as a flood or hurricane.

bedrock: the solid rock that underlies soil or loose surface material.

bituminous coal: dark brown to black coal of intermediate rank, softer than anthracite, harder than lignite.

bolson: southwest desert term for a low area that has no stream drainage outlet.

bottomland: flat, low-lying land, usually a grassy area on a valley floor along a river.

brachiopod: marine clam-like organism, but in its own phylum. Common in Paleozoic time.

breccia: rock composed of angular, broken fragments.

bryozoan: 'moss-animals,' resemble corals, but individuals are much smaller than coral polyps; in their own phylum.

butte: isolated, small, flat-topped hill, usually an erosional remnant carved from flat-lying rocks.

calcareous: rocks or materials containing calcium carbonate.

calcite: a common rock-forming mineral: calcium carbonate. Principal mineral in limestone.

caldera: the collapsed crater of a volcano, formed because lava is removed from the magma chamber below.

caliche: white layer or crust of generally nodular calcium carbonate formed in deserts by evaporation of soil solutions.

caprock: the hard top layer of a mesa or plateau; the impervious layer of anhydrite or gypsum over a salt dome.

cephalopod: a class of marine mollusks with coiled or straight chambered shells; octopus, nautilus, squid, and cuttlefish are today's examples.

chalk: soft, white, earthy limestone of marine origin, formed mainly from the shells of floating micro-organisms.

chemical erosion: where rocks are worn away by natural chemical processes, especially by solution and solvent action of water.

chert: hard, very fine-grained rock composed of interlocking crystals of quartz. Commonly occurs as nodules or concretions in limestone.

Chinese wall: narrow ridge of hard volcanic dike that stands above the landscape, and reminiscent of the great wall of China.

cinnabar: the red to brown sulfide mineral of mercury.

clay swale: muddy, swampy depression, as between two beach ridges.

claystone: sedimentary rock of mud or clay; also mudstone, shale.

coastal plain: broad, low area marginal to the ocean shore, its strata are flat or slope gently seaward; usually a strip of recently prograded or emerged sea floor.

coastline: the boundary between sea and land.

cobble: a rock fragment larger than a pebble and smaller than a boulder; usually rounded.

collapse debris: loose rock material in a cave, commonly below a sinkhole.

collapse fold: rock that is bent, contorted or folded because of nearby collapse of rock into a sinkhole or cave.

colliding plates: where the rigid plates of the earth's lithosphere come together, driven by seafloor spreading.

column: in a cave, where a stalactite joins a stalagmite.

columnar joint: parallel columns, polygonal in cross-section formed as lava contracts as it crystallizes.

compaction: the process of making rock from sediment, as water is squeezed out, and pore space is reduced from the pressure of the overlying column of sediments.

concretion: a hard ball or odd-shaped mass of mineral matter, usually formed from the precipitation of some minor mineral in the host rock; forms around a nucleus such as a bone, shell, leaf, or fossil.

conglomerate: sedimentary rock equivalent to gravel; formed of pebbles, cobbles, or boulders.

continent: one of the earth's major land masses; includes dry land and continental shelves.

continental drift: the process where the continents move relative to each other, driven by seafloor spreading and mantle convection.

copse dune: small sand dune anchored by vegetation.

coral: marine organism that has a polyp body and secretes a calcareous skeleton; corals are main contributors to today's reefs.

coryphodon: extinct group of primitive hoofed mammals of rhinoceros size, common in Early Tertiary time.

crater: the round, rimmed basin at summit of a volcano; formed by collapse, explosive eruption, or rim build-up.

creep: slow downslope movement of rock and soil; or slow motion of sand grains pushed across surface by wind in sand dune areas.

crinkled bed: sedimentary rock layers with minute wrinkles; commonly formed in carbonate rocks by algal mats.

crinoid: fossil echinoderm animal, "sea-lillies," usually having a long stalk or stem topped by a platy flower-like body of tentacles.

cross-bedding: sedimentary layers deposited at an angle to the main, horizontal beds.

crust: the thin outer layer of the earth; including continental and oceanic crust.

crustal plate: one of the discrete pieces of the earth's crust; which may be partly continent and partly ocean crust.

crystalline: containing crystals, often used to describe igneous rocks, where crystals form from cooling magma.

cuesta: hill with a gentle slope on one side and a steep slope on the other; formed by a resistant layer in an area of gently dipping strata.

cycad: common naked-seed plants that grew in the Mesozoic.

debris apron: broad sloping surface at the base of mountains formed by coalescing alluvial fan.

deflation lag: pebbles left behind on a surface after the wind has blown away the smaller sand and silt particles.

deformation: the process of folding, faulting, compressing, or extending rocks by earth forces.

diatom: microscopic, single-cell plant with silica walls; lives in oceans and lakes.

differential erosion: selective wearing away of harder vs. softer rocks.

dike: elongate body of igneous rock that cuts across the bedding of the surrounding rock.

dimetrodon: fin-backed reptile common in the Permian.

dinosaur: special class of reptiles dominant during the Mesozoic.

dolomite: sedimentary rock, or mineral, composed of calcium - magnesium carbonate.

down-cutting: stream erosion in mainly a downward direction (as opposed to lateral erosion).

downwarp: subsidence of a regional area of the earth's crust, often in response to uplift of an adjacent mountain range.

drainage divide: the high area between two adjacent stream valleys.

drip curtain: or drapery, in a cave; a thin sinuous sheet of calcite dripstone.

dripstone: general term for calcite deposited by dripping water in caves; includes stalactites, stalagmites, etc.

echinoid: sea urchin group of animals; generally rounded, covered by calcareous plates.

entrenched river: river that cuts into bedrock, having inherited its course from a previous cycle of erosion on flatter topography.

erosion: the wearing away of the earth's landscape by natural forces of water, waves, ice, wind, tides.

erosion rill: small, nearly parallel channels, created on steep slopes by flowing water.

escarpment: long cliff formed by differential erosion or faulting; a scarp.

estuary: wide funnel-shaped tidal mouth of a river, where sea water and fresh water mix; formed mainly by sea level rise and drowning of river mouth.

evaporite mineral: mineral precipitated as a result of evaporation: such as halite (salt), anhydrite, and gypsum.

exfoliation: weathering process where scales of rock are peeled off the surface, often resulting in dome-shaped hills.

extrusive rock, extrusion: igneous rocks erupted to the earth's surface, such as volcanic ash and lava flows.

fault: fracture in rock where one side moves up, down, or sideways, or is offset, relative to the other side.

feldspar: a group of aluminum silicate minerals that are the most common of any mineral group, making up 60% of the earth's crust.

flaggy, flaggy beds: like yard flagstones; rocks formed of layers that easily split into flagstones; distinct beds a few inches thick.

flask: jar used to contain liquid mercury; and unit of measure of mercury, or 76 pounds.

flatiron: a plate of steeply dipping resistant rock on mountain flanks that weathers into the triangular shape of a clothes iron.

flint: dark-colored chert; commonly used to describe arrowhead material.

floodplain: flat area adjacent to a river channel, covered with water when the river overflows its channel banks during a flood.

flowstone: cave deposit formed by flowing water on cave wall or floor.

foraminifer: protozoan animals that secrete a calcite "shell" of one-to-many chambers; mainly marine dwellers.

forebeach: beach on the ocean side of a barrier island.

formation: a distinct, mappable body of rock; named for a local geographic marker and usually the main rock type, for example Austin Chalk. Fundamental stratigraphic unit.

fossil: any preserved evidence of past life.

fulminate of Mercury: explosive compound of mercury, as used in military shells.

fusulinid: a common foraminifer in Pennsylvanian and Permian time; resembles a grain of wheat.

gangplank: sloping wedge of alluvial sediment; especially along the east front of the Rocky Mountains.

gastropod: a mollusk group, mainly the snails.

glacier: mass of slow moving ice, formed by compaction and recrystallization of snow.

glauconite: earthy, green, iron-rich mica mineral, formed as small pellets in shallow marine water where sedimentation is very slow.

gneiss: banded metamorphic rock in which layers of course minerals alternate with layers of fine, flaky minerals.

graben: long, faulted, down-dropped block of rock between two adjacent elevated blocks.

granite: common igneous rock, cooled from a magma body; contains visible crystals of the minerals quartz, feldspar, and mica or hornblende.

gravel, gravel bar: loose accumulation of cobbles, pebbles, boulders; if cemented, it becomes conglomerate.

gravity fault: the line along which a body of rock has slumped down a steep face.

ground water: subsurface water contained in porous rocks.

gypsum: soft, white mineral; watery calcium sulfate; commonly formed by evaporation along with anhydrite and halite.

gypsum rosette: flower-shaped form of gypsum found in soils and tidal flats.

headward erosion: lengthening and cutting upstream of a stream or gully, by rainwash, gullying or slumping.

headwaters: the source of a stream or river.

heavy mineral: usually dark minerals, of high specific gravity that form minor constituents of sedimentary rocks.

helictite: twig-like or odd-shaped cave deposits.

hoodoo: differentially eroded columns or pinnacles of rock that resemble animals or creatures.

hornblende: most common dark silicate mineral of the amphibale group; constituent of granite, gneiss, schist.

horst, horst block: elongate block of uplifted crust bounded by faults, adjacent to down-dropped graben.

humic: dark, organic-rich material.

hurricane: a rotating mass of tropical air in which the wind velocity is 74 mph or greater.

hydrocarbons: all the petroleum-related organic compounds that make up natural gas, crude oil or solid bitumen.

Ice age: a glacial episode; the Pleistocene Epoch.

Ichthyosaur: Porpoise-like class of sea-dwelling reptiles that lived during Mesozoic time.

igneous: rocks that solidified from molten material or magma. One of the three main classes of rock - igneous, metamorphic, sedimentary.

Inoceramus : large plate-like clam common in Cretaceous period.

interbedded: beds lying between or alternating with beds of a different kind; such as lava flow 'interbedded' with sediments.

intrusion, intrusive rock: igneous rock emplaced into other, pre-existing rock.

ironstone: rock rich in iron.

iridium: an element in the platinum group. Common in meteorites, rare on earth.

joint: a fracture in rock, where the two sides are not offset.

karst: topography characterized by sinkholes, caves, and underground drainage.

laccolith: a mushroom-shaped igneous intrusion.

lagoon: shallow water between barrier island and mainland.

landslide: downslope movement of a rock mass.

laramide, Laramide orogeny: Rocky Mountain period of mountain building, from Cretaceous to Eocene time.

lateral accretion: outward or horizontal sedimentation; in streams, point bars accrete laterally.

laterite soil: highly weathered, red iron-rich soil characteristic of humid tropical regions.

lava: molten extrusive rock, generally from a volcano.

leaching: selective removal or dissolving out of soluble constituents of rock by percolating water.

lignite: brownish to black coal, between peat and bituminous coal in coal rank.

limestone: sedimentary rock primarily composed of the mineral calcite (calcium carbonate); commonly deposited in marine water.

limonite: yellow-brown hydrous iron oxide mineral; rust; often the weathering product of other iron-rich minerals.

lithosphere: rigid crust and uppermost mantle of the earth.

llanite: a rhyolite rock with phenocrysts of red feldspar and blue quartz.

longshore drift: movement of sand along the shore, driven mainly by oblique waves striking the coast.

magma: molten material which solidifies into igneous rocks.

mantle: the thick zone in the earth between the crust and the core.

marble: metamorphic rock composed of recrystallized calcite; metamorphosed limestone.

marl: loose, earthy deposits of clay and calcium carbonate.

marsupial: an order of mammals, such as kangaroos, whose young develop in a pouch.

mass wasting: downslope transport of a mass of soil or rock by creep, or rockfall, or rockslide.

mastodon: extinct elephant-like mammals common in North America in the Pleistocene epoch.

meander: series of loops, turns, or bends in the course of a river; formed as the river swings from side to side across its floodplain.

mechanical erosion: weathering in which rock is broken down by physical processes involving no chemical change.

Mercury: heavy silver-colored metallic element; the only metal that is liquid at ordinary temperatures.

mesa: a table mountain; flat top, steep sides, bigger than a butte.

metamorphic, metamorphosed: rock formed from pre-existing rock by changes in temperature or pressure; includes slate, phyllite, schist, and gneiss, in order of increasing metamorphic grade.

mica: group of platy minerals, such as biotite; common constituents of igneous and metamorphic rocks.

mollusk: great group of invertebrate animals that includes snails, clams, nautilus.

morphology: study of the shape and form of things, such as landforms or fossils.

mud flat: low areas around barrier islands, alternately covered and uncovered by tidal water.

mudstone: rock composed chiefly of mud (clay).

nephelinite: dark, fine-grained igneous rock, either extrusive or intrusive; like a basalt, but has no olivine or feldspar minerals.

nodular, nodule: small irregular knot, mass or lump, such as chert nodules in limestone.

normal fault: a high-angle fault sloping 45 to 90 degrees, where the upper, hanging wall has moved down relative to the lower foot wall; produced by lateral extension or pull-apart.

novaculite: dense, hard, light-colored, very fine-grained siliceous sedimentary rock, thought to be formed from the silica bodies of marine microorganisms, such as diatoms and sponge spicules; used as a whetstone.

olivine: dark green silicate mineral rich in iron and magnesium; widespread rock-forming mineral in basalt.

orogeny: process of mountain building.

outcrop: rocks exposed at the earth's surface.

overburden: whatever rock lies over the rock you are interested in; to a miner it is the rock that lies above the ore, to a "Precambrian geologist" it is the entire overlying sedimentary section.

oxbow lake: the abandoned, water-filled loop of a stream meander that looks like a U-shaped ox yoke.

oxidation: chemical process where oxygen is added to minerals or other compounds; weathering oxidizes minerals; burning wood is a type of oxidation.

paleontology: the study of life in past geologic time, based on fossils.

palisade: a picturesque rock cliff; commonly refers to cliffs of columnar basalt.

Pangaea: the supercontinent that existed from about 300 to 200 million years ago; it split apart to form the present continents by fragmentation and continental drift.

peat: partly carbonized plant remains; an early stage in coal development.

pebble: a small, roundish stone; in size between a pea and a tennis ball.

pedestal: a column of soft rock capped by a wider block of harder rock; produced by differential erosion.

pegmatite: an extremely coarse-grained igneous rock, usually in veins or dikes; most commonly granite in composition; from the last-to-cool watery portion of a batholith.

petrified wood: wood that has been fossilized by silica replacing the wood.

phenocryst: in igneous rocks, a large conspicuous crystal in a finer groundmass.

phyllite: a shiny metamorphic rock, between a slate and a schist.

Phytosaur: crocodile-like reptiles common in Triassic time.

piedmont: an area at the base of a mountain range, such as a piedmont glacier, or a piedmont terrace.

plankton: small, floating marine plants and animals.

plate: a discrete segment of the earth's lithosphere; crust and upper mantle; may include oceanic and continental crust.

plate tectonics: concept that the outer part of the earth is composed of a number of plates that move horizontally over geologic time, driven by upwelling of volcanic material at mid-ocean ridges; it causes earthquake activity where plates collide and one plate is subducted beneath another.

Plesiosaur: group of long-necked, four-paddled, short-bodied, sea-going reptiles that lived in Jurassic and Cretaceous time.

plug: small, vertical body of igneous rock that filled the conduit of a volcanic vent.

point bar: series of low, arcuate ridges of sand on the inside bend of a river meander loop.

potash: potassium-rich carbonate rock.

ptilodus: group of Early Tertiary, forest-dwelling, primitive mammals.

pumice: rock blown from a volcano, filled with so many holes it generally floats on water. Used as a light-weight aggregate and as polishing and cleaning stone.

pyroclast: fragment of rock blown from a volcano.

recharge area: where an aquifer or reservoir rock is at the surface and catches and absorbs rain, which eventually becomes aquifer water in the subsurface.

reef: a mound or ridge built along a coastline by lime-secreting organisms, such as corals.

remote sensing: methods of geologic investigation where the recording device is not in direct contact with the earth, usually refers to photography, radar, infrared methods.

reservoir (oil or gas): subsurface rock with sufficient connected holes to store commercial volumes of oil or gas.

reverse fault: like a thrust fault, but steeper than 45 degrees.

rhyolite: very fine-grained, extrusive igneous rock; same composition as granite.

ripple, ripple mark: series of small ridges of sand produced when wind or water moves sediment.

river bar: ridge-like pile of sand in the channel, along the banks, or at the mouth of a river.

sabre-toothed cat: a large extinct carnivorous cat with enormous stabbing canine teeth that lived in North America until a few thousand years ago.

Salt basin, salt flat: low area, salt-encrusted from wet and dry evaporation cycles.

salt dome: pinnacle of rock that has penetrated overlying sediment; common along the Gulf Coast, forming oil traps in the upturned sediment layers on the dome's top and flanks.

saltation: process where sand moves by skipping along a surface in short jumps or bounces.

sand dune: pile of sand heaped up by the wind.

sand sheet: large plain of wind-blown sand, lacking large dunes.

sandstone: sedimentary rock composed of sand-size particles.

Sauropod: group of long-necked, four-footed, vegetarian dinosaurs.

scarp: see escarpment.

schist: metamorphic rock having visible, parallel flakes or grains such as mica; splits easily into slabs; between a gneiss and phyllite in metamorphic grade.

sea floor spreading: idea that an area of the ocean crust spreads apart by upwelling of magma along mid-ocean ridges; spreading an inch or two per year, to provide the dynamic force for plate tectonics.

sedimentary rock: one of the three great groups of rocks (igneous, metamorphic, sedimentary); consolidated layers of sediment.

shale: rock formed of clay (mud); breaks into thin layers.

shelf: flat area on continent margin; can be dry land or covered by shallow ocean water, depending on sea level.

silica, siliceous: silicon dioxide, the most common chemical constituent on earth; component of chert, quartz .

silicified wood: see petrified wood.

sill: igneous rock body that intrudes parallel to the layers in the surrounding rock.

silt, siltstone: sediment or sedimentary rock composed of grains smaller than sand, but bigger than clay.

sinkhole - round depression or hole where water drains into limestone underground. Common in karst terrain.

slurry: a highly fluid mixture of water and fine particles.

soda straw: a tubular stalactite.

source rock (oil): organic-rich sedimentary rock that yields oil or gas when subjected to proper heat, pressure, and time.

speleologist: geologist who studies caves and cave deposits.

spheroidal weathering: form of chemical weathering where spherical shells of decayed rock split off to create rounded boulders.

sponge spicule: tiny silica needles that form the skeletal structure of sponges.

stalactite: dripstone that hangs from cave ceiling.

stalagmite: dripstone that builds upward from cave floor.

stegosaurus: four-footed dinosaur with bony plates standing along its backbone; common in Jurassic and Early Cretaceous time.

strandplain: a long shoreline area built seaward by a series of accumulated beach ridges.

strata; stratigraphy: sedimentary rock layers; study of sedimentary rocks.

subduction, subduction zone: where one upper crustal plate descends under another as the two plates collide.

swale: low, swampy area, as between two beach ridges.

syncline: a downward fold of rocks.

tailings, mine tailings: loose piles of processed ore or discarded waste rock from a mine.

talus: pile of rock fragments lying below a cliff; derived from weathering of the cliff.

tectonics: the large-scale forces that deform the outer part of the earth.

terrace: a flat bench above and next to a stream, which marks a former higher stream level.

thecodont: group of Permian to Triassic reptiles that were the ancestors to dinosaurs.

therapsid: group of Permian to Jurassic mammal-like reptiles that were the ancestors of mammals.

Theropod: the carnivorous, two-legged dinosaurs, such as Tyrannosaurus.

thrust-fault: low-angle (less than 45°) fault, where the upper or hanging wall has moved up relative to the lower or foot wall. Produced by lateral compression.

tidal flat: flat, barren or marshy area that is alternately covered and uncovered by the tide.

titanoides: group of Early Tertiary mammals; found as fossils in Big Bend National Park.

trachyte: fine-grained, light-colored extrusive igneous rock, with feldspar and dark minerals such as biotite and hornblende as the major components; little or no quartz.

transverse dune: a long ridge of sand oriented perpendicular to prevailing wind direction.

Triceratops: three-horned dinosaur that appeared in Late Cretaceous time.

Trilobite: joint-legged, marine animals with segmented bodies that resemble pill-bugs; relatives of scorpions, spiders, and insects; common during Paleozoic time.

tufa: calcareous rock deposited around hot springs or seeps.

tuff, tuffaceous: rock made up of volcanic ash, pyroclastic debris.

turbidity current: a current of dense water and sediment that flows rapidly down continental slopes, triggered by earth quakes or oversupply of sediments.

unconformity: a significant break or gap in the geologic record, caused by uplift and erosion, or lack of deposition.

uniformitarianism: fundamental idea that the geologic processes seen today operated in the same general way in the geologic past.

volcanic neck: the narrow vertical tube or rock-filled tube that extends below the opening at the summit of a volcano.

volcanism: the process of extruding magma — lava or ash — to the earth's surface.

washover, washover fan, storm washover: small delta built on the lagoon side of a barrier island by storm waves breaking over the barrier island.

water table: the top surface of the water-saturated part of an aquifer.

wind stress: force of the wind on water; it produces waves and currents.

zoned feldspar: feldspar crystal that has distinct concentric layers, indicating a change in composition of the magma while the crystal formed.

Index

410

416